A
TREASURED
LIFE

SURRENDERING TO THE SIREN'S SONG

SECOND EDITION

RANDY LATHROP

A Treasured Life

Second Edition

Published by:
Signum Ops, 435 Nora Ave., Merritt Island, Florida 32952

ISBN 9798536682395
Library of Congress Catalog-in-Publication data:
Lathrop, Randy
A Treasured Life

Cover Design by:
Scott Hamlin
scott.hamlin@icloud.com

Photos by:
Randy Lathrop unless noted otherwise.

"Now and then we had a hope that if we lived and were good, God would permit us to be Pirates."
— *Mark Twain*

Contents

Chapter 1
Beginnings

There are many tales of lost ships along Cape Canaveral, wrecked there in foul weather, or grounded on the shoals offshore. The shoals and reefs near the Florida coast have earned their reputation for the treasures they have claimed. Many of these wrecks are on the shoals just off of the Cape, but the entire stretch of coast ranging from The Marquesas in the south to Cape Canaveral in the north has devoured its share of flesh and wood. The lore of treasure lost surrounds the Cape like a fog. An old salt told me that, "Some say that whenever a ship is in trouble and going down, you can hear sounds like singing in the wind," he said. "The salvors say it's mermaids, calling the ships. Lots of people claim to have heard it. It's not like wind or anything. They say when you hear it, you have no choice but to follow it, and you end up on the shoals. A number of saved souls have sworn to it."

＊＊＊＊＊＊＊

The *Daytona Beach News Journal* story of August 16, 1990 read:

"Threatened with arrest, the leader of a treasure expedition here thumbed his nose at the state Wednesday and said he is protected by a federal court order. For the past week, Randy Lathrop, 35, has been leading a team aboard the salvage ship Endeavor at a wreck site just off the Canaveral National Seashore. According to various sources, there is at least one Spanish treasure ship and possibly as many as four at the site."

I had just picked up a copy from the paper machine at the

1

little marina tucked in the north side of Ponce Inlet. We had been operating out of Ponce Inlet for several days, docked close to the Critter Fleet on the north side. It was a good thing we docked where we did as the article continued, "State Assistant Attorney General Eric Taylor reported that park rangers, the Florida Marine Patrol, and rangers on helicopters are monitoring Lathrop's activities." He also told The *News-Journal* on Thursday that the state hasn't ruled out physical intervention by law enforcement personnel if the salvage crew is spotted excavating sand off the site.

I'd heard that authorities including the Marine Patrol were looking for us at Port Canaveral, but fortunately they were looking in the wrong place. The following day, August 17, Friday's *News Journal* story didn't carry any better news. The headline read, "State Readies Legal Broadside in Dispute Over Treasure Hunt". I continued reading the article which quoted the State Assistant Attorney General:

"He, (Lathrop) is digging 20 feet into the sand. This is mining. This is excavation. It is our land. These are our waters. We have legal sovereign right to it."

I put the paper down, sorry I had spent fifty cents to read a bunch of threats from another government mouthpiece. I closed my eyes, thinking back, trying to remember the path that lead me to where I was at today. I remembered when the mermaids first sang their song to me... my mind wandered back to 1974.

✷✷✷✷✷✷✷

The tourists gazed down on the clear sapphire-blue water boiling up from the white limestone opening below. The contrasting brown and green colors of the surrounding forest complimented the scene as only mother nature can. A pair of manatees swam into plain view below the observation deck. Suddenly and unexpectedly the silence was broken with; "Oh my God, look at the whales!" Everyone turned their gaze from the manatees to the loudmouth with the Jersey accent. Mike knew they were manatees, he just enjoyed making a scene and acting a fool. His thick Jersey accent made it more believable, and he had the crowd convinced he thought they were whales. He was a member of our college scuba diving club. Mike Brady, he had the biggest mouth out of New Jersey, but, unfiltered as he usually was, Mike could be very funny.

Mike had us all red-faced and in stitches that day in 1974 when our dive club visited Blue Springs State Park. The club was supported by our college, which supplied us with a gasoline credit card and two passenger vans that could haul all ten of us and our dive gear. The club's membership had a full spectrum of personalities and talents.

Some, in particular, come to mind. George, for example, who we all suspected of performing crazy science projects on behalf of his employer, Florida Power & Light. He was an avid cave diver and part-time Brevard Community College student. His parties were always fun... he had a bottle of nitrous oxide feeding a SCUBA regulator via a 10-foot hookah hose[1]. He looked a little like Frank Zappa while wearing his dive mask. His hair and his mustache made him a perfect Zappa clone. George made our cave diving lights and our safety reels. This was gear made by someone you trusted with your life who had the talent to produce it... hopefully.

Then there was Dave M. who was a "diver's diver" if there ever was one. One look at Dave's gear told the story: his strap buckles on his fins were wrapped with duct tape to keep safety line from snagging while cave diving and he had a wet suit that featured patches glued on top of patches, a testament to many close contacts with the environment. His SCUBA tanks looked like they had been dragged behind a car on some rocky road thanks to many collisions with cave ceilings in Florida and The Bahamas. Dave introduced me to the art of spearfishing under the bridges in the Indian and Banana Rivers at night with a pole spear and a flashlight. We'd shoot all the Sheepshead and Spadefish we could eat, and what was left over we'd sell for eighty-five cents a pound at the colored fish market in Cocoa on Route 520. Back then in 1974, the visibility around the causeways was usually around ten feet or so. Most times we also found plenty of Stone crab hiding at the base of the concrete pilings. Clayton's Crab Shack in Rockledge would pay up to a $1.75 for Stone crab claws. Can you imagine? A buck and seventy-five cents a pound! Not today. No way.

---

1   A hookah hose in this context is a long hose normally attached to a SCUBA 2nd stage regulator leading to a compressor at the surface.

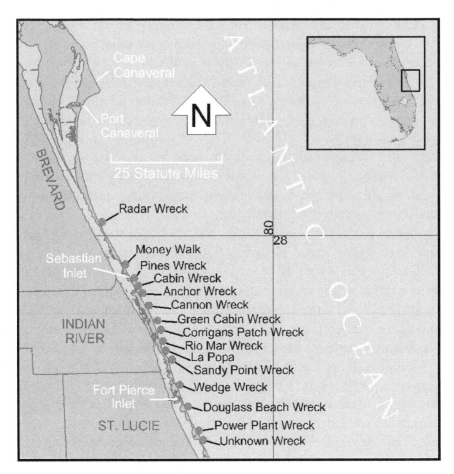

The area of Florida's east coast around Cape Canaveral is known as the "Space Coast", so named as the Kennedy Space Center is found there. South of Brevard County, the area of Indian River County and St. Lucie County is known as the "Treasure Coast", so named for the various shipwrecks of the 1715 Plate Fleet found along the shore.

The author lives in Brevard County. Divers who access the water from the beach only find acceptable underwater conditions south of Brevard County along the Treasure Coast where there are expanses of reef and rock bottom near shore.

I ate plenty of fried fish in those days. A two-dollar air fill, my Ikelight flashlight and a pole spear helped put me through college. I was a starving student, getting by on social security student benefits and 30 hours of working for minimum wage at the college library. When we weren't diving the springs of north Florida or the Florida Keys, we dove close to home. As long as we had water over our heads we didn't care where we dove. We dove into the quarries and dredge holes in the county, complete with gators.

We were ignorant of the possible dangers. I recall one dive along the shoreline in Fox Lake at Titusville, Florida, what I thought was a truck tire on the bank was, in reality, the tail of a huge gator. I slowly backed away, swam like a madman, dropped to the bottom of the lake far away from the bank, directly on top of another gator resting on the bottom. I didn't know whether to shit or go crazy! I imagine I looked like a Polaris missile coming straight up out of that water. All you could see was a trail of white water mixed with brown heading straight to the closest dry land. I curtailed my lake diving after that experience.

We frequently went diving from the beach at Sebastian and south from there into Indian River County. The Atlantic water cleared at Sebastian as opposed to points north of the Sebastian Inlet and a series of three reefs formed two-hundred yards offshore, with each reef separated by two-hundred yards more[2]. When we dove there, Dave M. showed us how to rig two large truck tire inner tubes together and secure two extra SCUBA tanks, your speargun, goody bag, and dive flag altogether, and off we would go. We would cover lots of ground with this method of crawl-

The author is seen here spearfishing around bridge pilings in the Indian River sometime in the '70s, when the water visibility was much better. The lagoon waters have degraded substantially over the last 50 years.

---

2   There is not much exposed reef or rock bottom along Florida's east coast from Sebastian Inlet northward. The Gulf Stream gets closer to Florida's shore toward the southern end of the state. Generally, the farther south you go, the better the visibility near shore.

Jimmy Ryan, left, and Terri Oldham are about to dive near shore using a canoe. This photo was taken in the vicinity of what we know today as the "Turtle Trail Crossing" at John's Island near the main pile of the Corrigan's Patch Wreck.

ing and pulling ourselves along the bottom of the reef. A pole spear works like an underwater ski pole. I could get going at a good pace using a pole spear to push me along the bottom. We dove this area often, especially during lobster season.

I remember the first time we came across one of the cannons from the wrecks of the 1715 Fleet[3]. The visibility in the area can range from a few feet to 30 feet but most of the time you can see less than 20 feet. The cannon came into view slowly, covered with seaweed and some marine growth. The sight captivated us instantly and my imagination was particularly piqued. I was told that the wrecks were worked by the Real Eight Corporation in the '60s and that most salvage had stopped for whatever reason. We saw no one in all our dives along the beach working any wreck in the early '70s. After we stumbled across that one cannon I never really thought too much about it until I read "Pieces of Eight" by Kip Wagner. I then read "Sea Fever" by Bob Marx. My interest increased.

3   The 1715 Spanish Treasure Fleet wrecked along Florida's east coast, and at least six of the wrecks are located throughout St. Lucie, Indian River, and Brevard Counties. It's believed that there were 11 ships lost. Not all of them have been found to date. There are dozens of cannons from these wrecks found near shore.

Dale Stone's Aqua Shack was conveniently located in the north-central area of Florida where so many springs and spring-fed rivers are found. Divers used his facilities to fill their SCUBA tanks and acquire special cave diving gear.

The dive club introduced to me lots of different dive sites across the state, but the springs and caves held much of our interest. It was from these trips that my fascination with cave diving started and forced me to become trained and certified in cave diving from NACD (National Association of Cave Divers) and the YMCA a few years later. These were the early to mid-1970s and cave diving was still in its infancy. We had new, improved, high-intensity lights, new buoyancy systems, new techniques, and Tom Mount's book "Basic Cave Diving" became our bible. We had three active certified cave divers in the club; George, Dennis, and Dave M. These three knew the areas very well and introduced us to many of the springs and sinkholes. We dove into the spring runs and lucked upon artifacts which once again drew me like a magnet to them. Bones, arrowheads, etc. We would stop at Dale Stone's Aqua Shack on Highway 27 Between Fort White and Branford to get our tanks filled. Dale Stone was a legend to those who knew of him and about him. Dale

The author is seen here fully dressed for a cave dive. Note the "horse-collar" buoyancy compensator, which was standard equipment in the early days of cave diving. Duct tape is used to wrap the strap fittings on his fins.

was one of the first people in that area to use SCUBA and dive the springs and rivers. *National Geographic* and many other publications featured pictures of Dale Stone and the relics and artifacts he recovered from the area's waters. His shop was a museum for many of his finds; one of Dale's better finds was the completely intact skull of a prehistoric sabertooth tiger. It was stunning. He donated most of his finds to museums. We can only hope they are on display.

Those were good times! The springs were yet to be commercialized, and many were unknown to most folks aside from a very few people who had permission to dive in them. We had permission to dive and policed them ourselves. Simple rules; no litter, lock the gate, don't frighten the livestock, and don't get killed! We frequented Ginny Springs, Devils Eye, Little River, Peacock Slough, Madison Blue, Telford, and Troy Springs.

I soon took a National Association Of Cave Divers (NACD) course for a week at Ginny Springs. My instructor was Mary Melton, and her husband at the time was Gene Melton, both being very accomplished and well-known cave diving instructors. Mary was an incredible cave diver. She was an excellent instructor, and I admired Mary for her knowledge and talent. I followed Mary through the woods with two 110 cu. ft. steel tanks on her back, lights, reels, and full cave regalia. Mary was not a big person. She was an attractive blond, of medium size, but a giant in cave diving. We spent a week learning proper technique, training, and diving as many of the local caves as we could. Most of our diving took place at Devils Eye, as I recall. Our class had maybe 6 to 8 students, and cost $1200, which was a lot of money in those days. But... how much is your life worth?

After the course, Mary's husband Gene, who was a pilot and a cave diving instructor, took us a few at a time for an aerial survey of the local springs and sinkholes. To say Gene was as enthused about flying as he was cave diving would be an understatement. Gene took up three people at a time, and when the first group landed, I knew I was in trouble. Most, if not all were green and gray upon exiting the small plane, grateful to feel the hot tarmac underneath them. My fears were further confirmed when I saw a charter captain out of West Palm, who sailed every day, looking like death warmed over as he got off the plane as fast as he could. My lunch was soon to be in the bag. The flight entailed a series of downward spirals for close-and-personal aerial inspection of interesting areas. I thought I was going to die. My eyes were probably closed 50% of the time when Gene would say, "let's take another closer look." I would start quietly praying. Upon exiting the plane my fellow students greeted me with anxious and evil anticipation of my ability to hold my lunch. I held up the brown bag with the contents of my lunch and yelled, "Fooled you, I still have my lunch!"

The club brings back fond memories. It was great until we backed one van into another car while fueling up on what would be our last trip. The college wouldn't sponsor us anymore. I continued to dive with many of the club members, cave diving, more technical, more intense. A favorite cave dive was "Eagles Nest", in Hernando County, it being what we referred to as a "shotgun dive", meaning you might encounter a pissed off landowner with a shotgun. The land on which Eagle's Nest sat was owned by a hunting club or possibly just leased by them. Over the years non-cave divers and a few legitimate cave divers would die at Eagles Nest, forcing the owners to take drastic measures. They dynamited two separate areas of the corduroy road which lead to the sinkhole. A corduroy road is an ancient road made of logs being placed side by side. We called these two deep dark muddy pits, "Little Mud" and "Big Mud". The first one, "Little Mud", was the first attempt at keeping the road closed. When that didn't work they tried again, henceforth, "Big Mud". It was a hundred feet of 2-to-3-feet-deep muddy slop. We used a 4-wheel-drive Jeep with winches front and back. We'd get out and tie off to a tree and winch ourselves across the mud holes. Eagle's Nest was, and is, one incredible dive. From the surface, it looks like a small farm pond maybe a few hundred feet in diameter. I remember there was a line tied to a log near the water's edge. The line would lead you to the center of the pond where there was a big hole that formed when the cave ceiling collapsed; it dropped 200 feet straight down. You'd fall through this hole in the ceiling at 50 feet and still be looking down at a huge sand mound 150 feet below. It was like dropping out of a hole in a ceiling into this large-size room almost the size of a large stadium. I remember when you passed 150 feet your exhaust bubbles would take on a metallic sound

The author is seen here on a cave dive at Little River Springs along the Suwannee River near Branford, Florida.

as they screamed upward for the surface. The darkness of the enormous room would just absorb your high-intensity cave diving light as if it were a penlight. At the bottom, around 210 feet, there were two separate caves. One was the spring side cave, with water coming out of it while the other, the "siphon" side, had the water flowing into it.

Two separate air supply rules always apply. When diving in a spring having outward flow, use the "one-third rule"; one third in, one third out, one-third emergency, and the siphon rule is; one quarter in, one quarter out, two-quarters for an emergency. A 20-minute dive at Eagles Nest would cost you several hours decompression hanging stationary on a line playing Tic-Tac-Toe on your dive slate with your buddy.

We used to sail down Highway 50 to Eagles Nest in Hernando County. It would take three times longer now. I still run into George on rare occasions. Dennis is a lawyer, and Mike B. was a commercial diver in the middle east before retiring to Thailand. I don't know what became of Dave M. — I can still see his smile. I can still hear Mike B.'s loud Jersey accent at Blue Springs, yelling, "Oh my God, look at the whales!"

Chapter 2
Diveco Diving Systems

I first met Bill Meyer in 1974, when I signed up for one of his dive classes at BCC (Brevard Community College, now Eastern Florida State College). He and his girlfriend Jan operated a dive shop in Vero Beach called, "Diveco Diving Systems". He taught several levels of SCUBA certification, from basic to assistant instructor at BCC and IRCC (Indian River Community College). Bill was a tough instructor. These were college-level classes with no promise that you would earn a certificate of completion. Jan was a little redhead that everyone liked, and wanted to know. Bill and Jan held the classes at the Cocoa Beach High pool because the college didn't have their pool at that time. Bill drove a cargo van with the coolest, very detailed, and colorful, underwater mural with divers painted on the sides. That van always smelled like soggy wet suits and dive gear, with 50 SCUBA tanks clinking together in the back. But I loved riding in that thing, watching the people admire the van as we sped down the road. People asked about diving at every stop, wanting to know this and that. It was always fun, even if it smelled just horrible inside.

Bill incorporated many tasks into his training regimen, some very creative. He would take eight tanks and lay them on the bottom of the pool, 10 feet apart from one another. Something was wrong with each tank; maybe it wasn't on, or maybe the baffles in the second stage didn't work, which forced you to use your hands to seal, or maybe it was just a tank with nothing on it, forcing you bare-valve-breath off the bottle. You had to take a breath out of each bottle to proceed across the bottom and complete the circuit.

He would black-out our face mask with tape, so we were blind, then have us follow a line on the bottom through different obstacles. It was fun to watch the students try to complete the course blind. Bill would bring different gear for us to try out including the old double-hose lung-busters from the Lloyd Bridges era[1]. These would fill with water when you inverted, and you had to draw hard and suck to get that single-stage regulator to let go of some air for you. You had to know your Navy decompression tables forward and backward if you were to pass any of Bill's classes. We would spend entire class periods going over decompression problems. Compared to what they teach now, it was a master's class. Nowadays, students just need to buy a dive computer to satisfy the class requirements. The curriculum mandates it! How do you plan a day's worth of diving for multiple divers with a decompression computer? Computers are wonderful, but being able to rely on the analog paper method to compute your dive is, in my mind, mandatory. Just plain old compressed air was also the only gas on the menus in those days. No Nitrox or technical diving, yet.

Bill and I soon became good friends. He brought me up to divemaster level so I could help with the classes, which I did for two years. I also got paid as a student assistant, which sure beat working in the library. Bill taught at two colleges, Cocoa and Fort Pierce. He also taught private classes at his shop on US 1 in Vero Beach. Helping Bill out with the classes was all kinds of good times... girls in swimsuits requiring a helping hand, you betcha! Bill talked me in to going for my instructor certification and had his girlfriend Jan and myself enrolled for a week-long course in Key Largo put on by Scuba Schools International (SSI). We stayed at Largo Lodge in Key Largo. A very wonderful throwback to the '60s, shrouded by multicolored croton plants on the bay side of US 1. A quirky property of six cottages as separate little bungalows, surrounded by tropical vegetation, with caged parrots hanging in the trees, little sandy footpaths going from one to another. Straight out of a Bogart movie, a lovely piece of history. I'm so happy to say it's still there, now called Largo Resort. However, I have not been there for over 20 years at

---

1   Lloyd Bridges was the star of the popular "Sea Hunt" TV series. He used the U.S. Divers Royal Master regulator based on an original design made famous by Jacques Cousteau.

least. Jan and I got through the course with the help of Bill's constant mentoring. I still remember Bill making me recite the four sinuses of the head, over and over until I repeated it like a rhyme. There are four paranasal sinuses, each corresponding with the respective bone for which it is named; maxillary, ethmoid, sphenoid, and frontal. I repeated, maxillary, ethmoid, sphenoid, frontal, maxillary, ethmoid, sphenoid, frontal, over and over. I still can name them today.

Bill and I dove often, and everywhere from North Florida cave systems to the Gulf, to the Florida Keys, and the wrecks of West Palm Beach. He introduced me to the Peacock Slough system and Orange Grove sink. This system is 28,000 feet of cave passageway. We did many "traverses", entering one sinkhole entrance and exiting out of another. Now the area is named after a friend, Wes Skiles, as a state park. That's another story I'll get to later.

We dove just about any place we could squeeze into, and one in particular with almost deadly results. We were in Hernando County, and this county accounts for a quarter of all the sinkholes in the entire state of Florida. We dove into Hospital Hole that day near Weeki Wachee spring. I've heard of two reasons for the ominous name. One legend has it that fish would come here to heal from the ocean, and the other was if you dove here you had a good chance of going to the hospital. The hole itself is in the third bend south of Roger's Park on the Weeki Wachee River. The main hole is about 150 feet in diameter and reaches a depth of 135-140 feet. Between 70 and 80 feet, there is a thick cloud-like layer of hydrogen sulfide that blocks most light below it. Below the layer, whose depth fluctuates with the tide, visibility is very good. The hydrogen sulfide layer — let me just say, "YUCK!" It tastes and smells like rotten eggs going through this layer of about 70 feet. You almost gag through this layer of a wispy rotten noxious film like stuff. And it will eat through a wetsuit if not rinsed off after the dive.

Later that day we ventured off into the woods at another location to an unknown sinkhole we named "Coffee Pot." The reason being that during the dive we found an old coffee pot, ancient and wedged in the bottom, so Coffee Pot it became. The cave had a small opening, more crevice than round, no flow what-so-ever. Bill went first as the reel-man. I must mention now that Bill and I were not cave certified divers at the time. I followed Bill. It was tight and silty. We descended to a small bottom at maybe 40-feet where we found the coffee pot. I took a quick look around. We realized that the silt was increasing, and we turned around. Neither of us could see anything now because the silt was so bad. Somehow Bill got in front of me with the line, a deadly move. I was off the line now, my only real pathway to the surface — the only way out of the darkness was that #24 braided line. I continued up the narrow crevice praying I was on the right track when Bill's fin appeared through the silt, I grabbed it and tugged at it so he would know I was there. We continued with Bill ahead of me as I held onto Bill's fin as we continued crawling our way through the passage with zero visibility. I had the coffee pot with me for a while but dropped it as soon as I began concentrating on getting out alive. Momentarily the darkness began to fade as we emerged from the tiny hole. We said nothing. We had screwed up, and it almost cost me my life. A reel-man is first in and last out. Somehow we had violated this rule. Bill had gotten ahead of me, and that shouldn't have happened. He should not have exited until he knew I was on the line ahead of him. We drove home that day in silence, reflecting on what had just happened and how our lives could have changed that day. After that day, I did the right thing and invested in my education and my life. I took an NACD Cave Diver Course, got cave-certified, and learned to dive caves safely and with other cave-certified divers only.

I continued to help Bill and Jan doing a few classes for them and helping with open water dives. We drifted apart some but always kept in touch. Then one day in 1979 the news came that Bill was dead. Bill was killed in a plane crash in the jungles of Colombia, not much more information, and I'll leave it at that. I heard his family was having a troublesome time getting his body back to the states. I don't know what happened to Bill's body. The shop sold, and Jan soon disappeared into thin air as well... I know little more. Rumors were whispered that Jan and Bill were together in South America. The story goes they were in debt, and it was all a comprehensive plan to escape to a new life. I fantasize about waking up in a beach bar somewhere down south in the Caribbean, and there sits my friend

Bill with Jan at his side. But it's just a fantasy. The tears come with the memories of Bill and Jan. That smartass grin of his and the way his bangs curled up, how he hated eggs over easy... he couldn't even watch you eat them. Miss you guys. Love ya.

Chapter 3
Pull Over Here

The yellow Toyota had served its first owner well and now was serving a second tour with its new owner who had painted it with the canary yellow house paint he had remaining from his last job. It was a mixture of rusted metal and fiberglass fused into one shape to cover the holes which had resulted from being parked on the beach for the last five years. The yellow would soon turn to gold.

As the sun crept over the horizon it passed by the rockets standing tall along A1A in front of Patrick AFB. The houses along this stretch of coastal Florida were remnants from the late '50s and early '60s. They housed the families whose fathers worked for the many space contractors which had swarmed in during the early days of the space program. They were small cinder block homes, with a surfboard or two propped up along one side. The Toyota pulled into the McDonald's as the three occupants shouted their orders into the squeaky speaker. Jim, the driver and owner of the yellow Toyota, asked the other two, "What do you guys want?"

Alex answered, "I'll have an Egg McMuffin." Alex owned the Toyota originally before passing it along to Jim.

"How about you Randy?"

"Make that two, I think I have enough money."

It was June 19th, 1977, the beginning of summer and I was looking for work as I simply existed from job to job. Most of the time I was only two weeks away from being on the street. At the time it didn't seem to sink in, or I just didn't understand how cruel the world might be if you let it. We were always happy and as my mother once said, "Just as happy as if we had good sense". That adage seemed to be more relevant to me as I grew older.

As the wrappings came off the breakfast sandwiches, the car continued down A1A with the three of us eating and washing it down with orange juice. The 80 cubic foot SCUBA tanks rattled in the trunk as the car hit a pothole here and there. We drove by rows of Australian Pines brought to the state 100 years ago to provide shade and windbreaks, eventually crossing over the bridge at Sebastian Inlet, which connects one barrier island with another, a point where the Sebastian Inlet links the Indian River with the Atlantic ocean. The bridge is close to where survivors of a Spanish Fleet shipwrecked in 1715 camped for a short time while they attempted to recover their lost riches[1]. Other salvors would occupy the same camp for many years to come. The site, more popularly known as the "Cabin Wreck", was added to the U. S. National Register of Historic Places on August 12, 1970. (See map Page 4)

We continued south on A1A to the vicinity of John's Island, a gated community inhabited by the rich and elite driven away from their original enclaves, now overcrowded in West Palm and Fort Lauderdale. The car cut off the road just north of the first security gate and headed down a sandy trail through the surrounding jungle at a quick pace so as not to get stuck in the loose sand. As soon as the car stopped, it was a contest to see who could get to the edge of the dune and see what the ocean looked like that morning. It was here that James Larouch homesteaded land in 1884. How different it must have been then from how it is today.

On this particular day, it was flat calm and the water clarity was incredible! There would be exceptional visibility underwater, a boon for divers as water clarity was not usually so good along this stretch of shoreline. This morning we might be enjoying the crystal clear waters of the Florida Keys or The Bahamas. The water was turquoise blue with visible dark patches of the reef shining up from the bottom through the remarkably clear water. We rushed back to the car and dragged our gear down to the wa-

---

1   This salvage camp was located near present-day McLarty Museum within the boundary of Florida's Sebastian Inlet State Park.

ter's edge in as few trips as possible. We didn't have much money, but we could usually afford an air-fill for two bucks if that might translate into a few pounds of seafood in our empty refrigerators.

Jim and I had taken our diving classes together at Brevard Community College, and we dove together frequently. Alex was a surfer and waterman but SCUBA was new to him so he was under our watchful eye until he became certified sometime later. I ended up certifying him as a matter of fact.

"Damn, I can't believe how clear the water looks," I said.

"Come on you guys, quit goofing around, let's go," was Jim's answer.

Alex looked serious as he sorted through the gear and glanced one more time at the ocean. Soon all three of us suited up with Jim leading the way into the ankle-high surf, with me following, and Alex cautiously bringing up the rear. We waded through the surf pulling our masks down over our faces as the water became deeper, finally placing our snorkels in our mouths as we paddled out from the shore. As we swam, you could hear the reef cracking and popping reminding us that it too was a living breathing thing. The bottom looked close through the crystal clear water, and we rarely witnessed such clarity in this area, but today you could easily see 50 to 100 feet in front of you. The bottom took on a whole fresh perspective, it reminded me of a green meadow with small flowing hills of rock covered with a fine green moss. The shafts of sunlight poured down through the water making it look like the background for a religious painting of some sort, perhaps exalting the second coming of Christ.

We swam east towards deeper water, Jim in the lead, myself second, and Alex tagging slowly behind. We swam over ledges of rock and potholes, peering down to spot an occasional lobster staring back at us from the security of the reef. Once we were in 12 feet of water, we headed north and swam some distance, then slowly turning our way back to the west, and then slowly swam south to complete our circuit. As we swam south Jim advanced his lead wanting to be the first to see what might lie around the corner. I swam along, peering around, occasionally glancing back to make sure Alex was still there. The reef popped and cracked under us as the shafts of sunlight shot down through the water and reflected off the green moss-covered rock. It was a beautiful day. Jim was way out in front now and when I turned to check on Alex... he wasn't behind me anymore! I turned and retraced our route looking for Alex. I could just barely make him out on his knees reaching into a pothole in the reef. As I swam to him, his arm pulled back from the pothole and he was holding a round golden object in his hand. I came closer looking at Alex. His eyes were wide and questioning as he held out the object for my inspection. It appeared to be an oval-shaped platter made of gold and decorated throughout and on the edges with small sunflowers. It was beautiful, and it had hardly any marine growth on it at all! It was heavy, and something very, very, special! I motioned for him to surface. As we broke the surface I fell into a diatribe of obscenities.

"Alex, Alex, holy shit!"

He looked at me as if to say, "What are you talking about?"

"Alex, it's gold! Alex, it's gold!" Alex hoped that was the case but just found it too unbelievable. About this time Jim had retraced our route and surfaced, swimming over to see what the fuss was all about. Alex handed the tray to Jim, who took it quickly and turned it over and over in his hands. He handed it back to Alex and told us both to calm down as we looked toward the shoreline to see if anyone was watching us. Alex took the tray and slipped it inside his wetsuit, and we cautiously eyed the shoreline for anybody who might have seen us.

We submerged and tried to relocate the same hole where Alex had pulled the tray from, as we had drifted off the spot in our excitement. Swimming in circles around the area, we were never entirely sure we had located the same hole. Finally, it was time to swim ashore and examine our new find more closely. Once ashore we raced back to the car parked back on the dune where we tore off our gear, throwing the wet sandy pile of chrome and neoprene back in the trunk. Jim pulled the car back onto A1A and we took turns looking at the "plate" as we now called it, passing it back and forth and wondering how much it was worth? We came to the stoplight at Wabasso Beach, where there is a little convenience store that still stands there to this day. We bought a six-pack of Busch beer tallboys to celebrate. As the alcohol took effect, we became more cavalier, showing off our new find to passing mo-

The author is seen here displaying the gold tray, or "plate" as it was originally annointed.   Photo: Jim Ryan

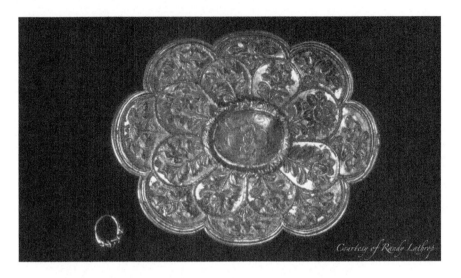

The gold tray prior to being cleaned. The ring at lower left is in place for scale.
Photo: Alex Kuze

torists. As a car would pull up next to us at a stoplight, we would hold up the gold tray, showing it off to the unsuspecting drivers next to us.

In short order, we arrived at Alex's apartment, where Alex hid the tray in a "secure place". The secure place according to Alex was at the bottom of a pile of his dirty clothes in his closet, which seemed good enough. I had read "Pieces of Eight" by Kip Wagner, who was the first to rediscover the wrecks in modern times, so I knew his company, the Real Eight Company, had a museum in Cape Canaveral. I suggested we drive to the museum and maybe try to get an idea of what the tray was worth. At that time the Real Eight museum was on its way out after having been broken into. It seems the company of original treasure hunters were now going their separate ways after their big finds in the early '60s. We walked into the museum with a totally different outlook on the treasure hunting business that afternoon than we might have had earlier in the morning. I do recall that our hair was still wet! They took our money and handed us the little tape players which would narrate the exhibits as we roamed through the museum. We were wide-eyed as we slowly walked from exhibit to exhibit hanging on every word. I had always thought they salvaged ALL the treasure in the '60s and did not understand that any remained. I was definitely wrong! Finally, as we came to the end of our tour, came the recorded words I'll never forget from the museum tape-player. The recording thanked us for visiting and hoped that we might tell our friends to come and to not forget that treasure remained just off our coast. The closing remark was something along the line of, "... who knows, today could be your lucky day!" As the tape ended we all looked at one another and started laughing.

The Real Eight Company museum is seen here, circa 1972. The museum was located in Cape Canaveral, Florida on A1A. Photo courtesy FloridaMemory.com

Chapter 4
Here Come The Sharks

After a few weeks passed, we tried to figure out what to do with the "plate". Jim said he knew an attorney. He said he had met him while shucking oysters at one of his employer's parties. His employer just happened to be the mayor of Cocoa Beach. We collectively figured that an attorney would be a good first stop in case we had unknowingly broken any laws. We were trusting our future to an oyster-slurping lawyer Jimmy had met while shucking oysters at a party. What could go wrong? Jim made the call and set up the appointment at the lawyer's office in downtown Cocoa Beach. After the Apollo program ended and before the shuttle program was in full swing, the town suffered from layoffs. Most families lost their jobs, could not sell their homes, and some just walked away from them. It was now a sleepy little surfer's hamlet with a handful of space workers and retirees sprinkled about the community.

When we met at the lawyer's office, Alex brought the gold tray with him in a paper grocery bag, double bags mind you. The receptionist showed us into the attorney's office.

"Hi, I'm Walter," he said. "How can I help you?"

Jim started telling our story with Alex and me jumping in when we felt required to do so. Alex opened the bag and took out the gold tray. Walter's eyes became fixed on the tray as Alex handed it to him. It's hard to appreciate the feel and vibe of the tray unless you've ever held something of such value both monetarily and historically. Walter remained quiet, turning the tray slowly over and inspecting it.

"Wow," he whispered. "So how much do you guys think this thing might be worth?"

We all thought for a minute and then I blurted out, "Hell, it's got to be worth at least $25,000.00!"

Jim and Alex both agreed, $25,000 was a lot of money! I didn't know it then, but I had become an expert appraiser on early colonial Spanish antiquities, as that appraisal blurted from my lips that moment would stand as we would find out. Walter called in his partner and his receptionist to come into his office. He handed them the tray so they too could feel the antiquity flow from the gold piece. The lawyer suggested that it be placed in a safe deposit box at the bank.

I need to mention that during the time we waited to see the attorney, we made many trips back to the treasure beach to see what we might find. We located two cannons in the surf and on another occasion we found a twenty-dollar bill stuck to a sea urchin! We had that twenty changed at the Wabasso convenience store into $6.66 to split three ways... all for one and one for all.

On another occasion, it was late towards the evening and I was the last one coming in from my dive when I surfaced to see where I was. I saw Jim and Alex waving at me frantically from the beach. They were waving and pointing at something. I submerged and made my way along the bottom which was easier than swimming on the surface. Visibility was not over 10 feet, which was normal for this area, not the super-clear water we had when we found the tray. I would go along and surface once again to check and see how far I was from the beach. Each time I surfaced Jim and Alex would wave and be pointing. As I surfaced my last time in the surf, Jim and Alex came running down and grabbed me and drug me up the beach.

"What the hell is wrong with you two?" I sputtered.

"Dude, you didn't see all the surrounding sharks?"

"I saw nothing," I replied.

"They surrounded you and we thought you were a goner, no shit," said Jim. Alex chimed in, "There must have been ten sharks around you at one time... one was an easy 10 foot!"

I was stunned: "You shitting me?"

"No, we thought you were fish food, especially the way you would come up and then go back down again," said Jim.

That was one bad thing about cloudy water; you couldn't see all the sharks. Ignorance can be bliss at moments like these, you can't fear something you can't see I guess.

The days passed and we dove and researched, trying to find out all we could about the 1715 fleet and treasure. Walter called all of us into his office one day and shared the news. We had made one of the worst decisions of our lives when we had notified the state of our find. After a few phone calls to the Department of Historical Resources in Tallahassee, a meeting with the State Underwater Archaeologist was scheduled in Walter's office. We arrived at Walter's office, where we were introduced to the State Archaeologist. He looked like a white version of Billy Preston with a rather large afro, glasses — maybe Ward Cleaver on acid. His name was "Sonny". Jim and I introduced one another, while Alex and Walter went to the bank after the tray. Alex returned with the same paper bag it had always been in, and he handed it to Sonny. Sonny opened the bag and exhaled slowly. He remained quiet not saying a word and he handed the tray to his associate, who forced an equally long exhalation. Then came the questions; "When did you find it?", "Where did you find it?", "How did you find it?", "Is there any more?" We told our story once again finally allowing Sonny to talk. Sonny informed us we had broken the law and by all rights, he could take the object with him now but would settle on leaving it in the deposit box. However, he insisted that both he and Alex were to be present whenever the box was to be opened in the future. He also informed us that the submerged area where we found the tray was under lease to a company out of Pennsylvania. Sonny told us that although the cards were not in our favor, he would do what he could to reward our honesty.

After we took a series of photos, the tray was returned to the box at the bank. Sonny told us that the first person who may be entitled to the tray was the person leasing that stretch of coastline from the state for treasure salvage. We got his information and had Walter contact him to tell him of our find on his lease, hoping beyond hope to strike a deal with him instead of the state. I don't know exactly how the conversation went with the leaseholder of the site, the state, and our lawyer, but I was told it went something like this...

"Do you mean to tell me that three boys found a gold tray on my lease and you are calling me to tell me about it? I'm so sick and tired of those crooks down there. You send the papers to me and I'll sign it over to the boys!"

This guy was fed up with dealing with the state of Florida. Now, this might all be a lie when I look back on it. I don't recall ever seeing a document between us and anyone else. Maybe the state said nothing to anyone, especially the contracted leaseholder who would have full rights to the tray, making it very difficult for the state to possess as they might in dealing with us instead. Maybe our lawyer wasn't in on the call. Maybe there wasn't a call. Now the state had a clear path to deal with us, three boys who had unknowingly just found the largest gold artifact ever recovered on a Spanish galleon in Florida waters, or in United States waters for that matter! They just couldn't take it from us as that would create horrible publicity for the state, and no one would ever turn anything over to the state again if that happened and word got out.

Photo: Alex Kuze

Chapter 5
Coffee With The Duke

It was 1978 and almost a year had passed while the wheels of justice — or injustice in this case — turned slowly. Life went on for the three of us trying to get by the best we knew how when a call came in from a guy out of California looking for a few local divers with local knowledge. The man's name was Chuck Kenworthy and he was running his salvage operation out of a marina in Sebastian while living in a funny-looking roundhouse he rented on A1A near Melbourne Beach. He had two sons that would hang out with him in the summer and help. Chuck also had two nice boats. For those of us who had been using truck tire inner tubes as our dive vessels, this was a move in the right direction. We would have propellers and everything!

Chuck was aware of the gold tray find, which was the main reason we were now talking with him. He knew we had the fever and he could also surmise from the beat-up yellow Toyota we showed up in that we would not be difficult to deal with when it came to setting a wage. The bottom line — things were looking up, the boats were great, the bologna sandwiches were fresh — we were in heaven! Chuck had us hypnotized from the first meeting. This guy could sell an idea or a dream. If anyone of us would have had any money, we would have invested ourselves, he was that good. He was so good several of his California investors were big names. One, in particular, was a big fella his friends called "Duke". Chuck named his boats after Hollywood movies, one he named after one of Duke's movies. The main boat was a 42-foot Chris Craft with twin blowers, named, *Quo Vadis.* The second boat was the mag boat, and they called it the *Searchers*[1]. He even let it slip out that occasionally the big man comes east and hangs out on the boat for a few days now and then. Holy shit, we like to have died and gone to heaven when we heard that, hell, we'd almost work for free! Chuck could read our reactions and our minds. We sobered up and negotiated our price.

The captain of the *Quo Vadis* was the image of a storybook sea captain, small grey beard, weathered tan, always had an unlit pipe in his mouth, and was prone to sporadic outbursts of cantankerous grouchiness. This was to be our introduction to the techniques and mechanics of modern-day treasure hunting. We would come to love and thrive on such things as rocking and bucking work platforms, diesel fumes, wet swim trunks with perpetual sand in the crack of your butt. Captain Doug Dykes taught us the basics. He was a good mariner, he was safe, we learned something new every day, many times at the expense of our egos, but we learned.

The second boat, the *Searchers*, (named after one of John Wayne's movies) was around 26 feet long and pulled the mag for the remote sensing surveys. Fay Feild was in charge of this boat. Fay is credited with developing one of the first magnetometers used for treasure hunting applications. A magnetometer is a device that measures any difference in the earth's magnetic field which might be caused by ferrous metal. The story goes that Fay's former wife was an avid shell collector and one rare particular type of shell was found primarily near iron wrecks. So Fay developed the magnetometer to locate iron-hulled wrecks, hoping to locate this shell. Fay had worked with Mel Fisher for years and was contracting his services during slack periods caused by Mel's periodic financial difficulties.

One hot summer day Fay and I were pulling the mag on the Corrigan's wreck site using the smaller vessel, *Searchers*. Fay always had the ever-present, chewed-up cigar in his mouth as he watched the plotter hooked up to the mag record any hits we may have crossed over. This area can be shallow and has lots of hard rock on the bottom.

---

1 A magnetometer is towed behind a "mag" boat. Generally, mag boats are smaller vessels which can maneuver closer to shore, dodging waves and shallow reefs.

There's plenty of reefs in the area. In order to pull a mag across this terrain, the mag's "fish" must be modified, so it doesn't drag on the bottom. To accomplish this, we suspended the fish below a surfboard. The surfboard rode on the surface with the mag fish suspended just a few feet below it. We were pulling the mag at a few knots doing our survey and trying to stay awake, with Fay staring intently at the read-outs on the mag's plotter. The plotter needle was doing its thing going up and down when suddenly it just went "whoosh", and it was gone! The mag fish had snagged the top of a reef close to the surface and got

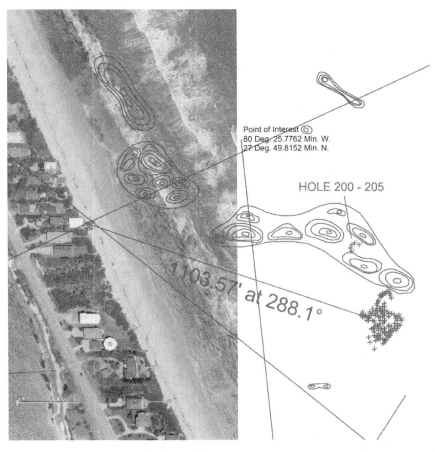

A magnetometer map produced by Doubloon Salvage in 1985 is seen here overlaid on an aerial of the Cabin Wreck just south of Sebastian Inlet. Note the intensity of registered iron anomalies seen as intensity gradiations.

hung up almost pulling the entire assembly overboard, cable, plotter, and all! Fay bit his cigar in half as the plotter went sailing towards the stern. I shut down the boat just in time before we lost everything. That woke us up!

Work started in May with the season running from May to September when the waters allowed for clear visibility and calm seas. The salvage season ends when the hurricane season would come to a head in September. But most summer days were grand. We would stand on the deck as the boat charged through the waves of Sebastian inlet. We were proud to be aboard. We always stood out on deck hoping to be seen by as many people as possible, proud of our association with such a grand endeavor. We had come a long way from diving via inner tubes. Many mornings we noticed that we shared the waters with an ugly, ragged, little scow of a boat no longer than 20 feet plying the turbulent waters of the inlet alongside us. In this sorry looking little tub was a guy with long red hair and a red beard to match. He smirked and would watch us intently as we separated from one another at the mouth of the inlet. We'd laugh and point at this pitiful excuse for a boat, having forgotten our most recent glamorous background of inner tube yachting only months prior. I'd seen this guy several times before, and also on the beaches. More to come on this fellow.

When we arrived at the area we were to excavate, the first order of business is to anchor the boat in a four-point mooring. We would do this by dropping four anchors, two spread out from the bow forward, and two spread out from the stern, anywhere from 50 yards to a hundred. Captain Doug was good... he would place the two forward anchors, then have the divers swim the aft anchors to where he wanted them dropped. This was done by placing the 44-pound Danforth anchors on.... you guessed it, a truck tire inner tube. Old habits die hard and the past always comes back to haunt you. Captain Doug would stand on deck, pulling a white hanky from his pocket and waving it to guide us to the correct spot to dump the anchor off the inner tube. The image of him waving that hanky at us like some weeping bride waving goodbye to her sailor boy as he leaves for the sea sticks with me. Once the anchors were in place, we would winch ourselves and maneuver where we needed to be according to Chuck's fancy aerial view chart with all the adhesive dots showing where holes had been blown previously in the ongoing search effort. Once the boat was in the desired spot, the next move was to

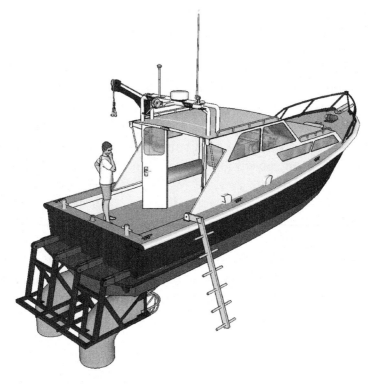

A contemporary treasure dig boat, the El Salvador is shown here with blowers in their "pinned" position.   Illustration: T. L. Armstrong

lower the blowers and pin them into place under the hull of the boat. Blowers are tubes that fit over the prop and deflect water towards the bottom, blowing away the overburden. Pinning the blowers almost always required two divers; even the smallest groundswells and waves made this procedure dangerous. It would be very easy to lose a finger or a hand with the blowers rocking in the murky water.

With the blowers now pinned, and the boat anchored properly, the engines were placed in gear turning the props, throwing propeller wash towards the bottom, blowing away the sand and overburden, and excavating a hole, hopefully exposing wreckage or treasure at the bottom which had laid there for hundreds of years. While the prop wash was doing its job in deep sand, we usually had a half-hour before our services were needed as divers to investigate the newly excavated crater on the seafloor. The hole would simply fill in almost immediately, so the engines had to be running just enough to keep the hole open for the divers to make their inspection. During these interludes between excavations, we would rummage around the cabin looking at magazines and drinking coffee from a special coffee mug. Onboard the *Quo Vadis* there were a collection of coffee mugs designed specifically for Chuck and his guests. These mugs were molded plastic with a picture of Chuck and the Duke on one side and a fake gold doubloon embedded in the other side of the mug. Also, there were names on the cups to establish whose mug it was. Well, on one mug there was the name "DUKE". John Wayne's coffee mug... wow! In the days that followed, there were frequent squabbles about who got to drink out of the Duke's coffee cup, which was the "Holy Grail". Holding the Duke's coffee cup in your hand, looking out at the horizon, bouncing over the waves, sharing spit with the Duke... life was good! The days went by and our knowledge grew.

We became proficient at anchoring, pinning the blowers[2], diving the craters, and living with sugar sand in all the wrong places. We spent most of our time north of Sebastian Inlet at a site designated as E56, meaning "Exploratory permit 56". We hated E56. It was deep sand, took a long time to dig, and the bottom was virtually barren everywhere you looked; it seemed to be a desert.

Chuck had the leases on two sites at this time; the previously mentioned E56, and S19, which was Salvage permit 19 located south of the inlet, more commonly known as "Corrigan's". (See map Page 4) When the boat would turn south instead of north, our spirits would

---

2   Propwash deflectors, also known as "mailboxes", or "blowers" must be placed behind the boat's propellers and fixed into position. Blowers are usually attached to an armature that is hinged to the boat's transom. Once the blower(s) is pivoted into position, it's customary to have some linkage from the blower to the propeller's struts using a pin of some sort to bind the two to one another.

immediately soar, and our anticipation grew with every mile as we got closer to Corrigan's. We knew there was treasure there having found it there before ourselves! On the days we'd turn to go south of the inlet, often we would once again pass our red-bearded friend in his little scow, leaving him in our wake. Our red-bearded friend turned out to be a fellow by the name of John Brandon from Fort Pierce. John and I eventually would become good friends and we would work together in the future. We found out John was working an early wreck, which sank in 1618. This is known locally as the "Green Cabin Wreck", so named for the green cabin that stood nearby on the beach[3]. The wreck was identified at some point as the Honduran Almirante, being one of the major ships in a Spanish plate fleet. One day as we passed "Red Beard", he had all kinds of buoys, marker floats, and yellow lines spread all over the Green Cabin Wreck site.

The Corrigan site was suspected to be the admiral ship of Captain-General Don Antonio de Echeverz y Zubiza and the ship was named *Nuestra Senora de Carmen y San Antonio*. It was one of twelve ships, eleven of which sank in 1715, settling on the bottom along the Florida east coast from Fort Pierce to Sebastian inlet. The bottom here differed greatly from the barren sands of the E56 site. The bottom here was alive with exposed reef rock, marine life, and shallow sands covering the bedrock. We would see a variety of fish and marine fauna, along with many dorsal fins of "the man in the gray suit" cutting the surface of the water about every day. It was less difficult to remove the overburden in this area, as the sand covering the rock in this area was quite shallow compared to the sands of E56. It only took us 10 or 15 minutes blowing before we dove for a look. As I mentioned previously, Chuck had some high-quality aerial charts we used by applying a little red or green sticky dot on the chart where we had excavated. A red dot meant nothing was found and a green dot stood for shipwreck material found. We soon had the chart covered with dots, mostly red ones but just a few green ones. We found scattered "pieces of eight" and assorted pottery shards. The underwater metal detectors in the '70s were nothing compared to

---

3   Wrecks from different eras are co-mingled along Florida's coast, some
     being found atop others in the same location. Wreckage from a 1715 ship is
     interspersed with the Green Cabin Wreck.

There are thousands of pieces of "bojitas" olive jar shards found strewn throughout the 1715 Plate Fleet wrecks. This shard plainly shows the finger grooves of the potter's hand, left in the jar as it was thrown on a wheel, 300 years ago.

today's underwater metal detectors. Often they weren't working, so we eyeballed it most of the time, looking for the blackened, greenish oxide encapsulated coins mixed in with the rubble, swimming from the bottom of the hole in a spiral towards the top, carefully surveying the sand, shell, and rock. Who knows how many coins we left behind, just waiting for the next salvor with the right technology?

We would work Corrigan's for a few days, then head back north of Sebastian to the deep sands of E56. We couldn't understand Chuck's fascination with E56. We figured that Chuck regarded Corrigan's (S19) as money in the bank, while he continued to explore E56. Why was Chuck so fascinated with this area where there were no cannon, no ballast, no pottery, nor anything on the bottom to indicate a shipwreck was in this area? We came to discover that Chuck had an agreement with a beachcomber who lived nearby.

These are Eight Reales, or "pieces of eight", recovered from the 1715 Plate Fleet wrecks. They are also known as "cobs". Cobs were made by hand. Pieces of eight weighed about 27 grams, and were about the size of a silver dollar. The cob on the left has been cleaned while the cob on the right is seen as found. Note its sulfided surface . There's also some rust on this cob, indicating that it laid close to some iron. The clean specimen shows the mint mark of the Mexico City mint as "o" over "M", and below that is a "J", denoting the assayer, Jose Estaquio de Leon. Most of the cobs onboard the 1715 Fleet bear his assay mark.

This beachcomber had shown Chuck the multitude of Spanish coins that he would find after storms. As an aside, I now know this person very well. We met and coin hunted a little together much later, but that's another story I'll get to soon enough. Most of the coins were 4 reales and 8 reales in remarkable condition. I would find out for myself why Chuck was fascinated with this stretch of beach [4].

The summer burned itself out and the days we could get out and work were becoming fewer. We continued to dive both sites and shared the summer dive season with two state agents who were required to be on board when we were on either site. They were mostly interested in any recent issues of Playboy or Hustler that might be onboard. On a side note: *Florida Today* newspaper did a small article, titled, "Sea ties Wayne to Brevard", about the Duke's treasure fever. The winter would soon be upon us and there would be no more diving. But the good news was the state was ready to make us an offer on the plate.

---

4   There was a popular steakhouse in this area near Brevard County's
     Bonsteel Park, more popularly known to beach combers as "Money Walk". It
     was not owned by Kenworthy, but it was known as "Chuck's" steakhouse.

Chapter 6
Tallahassee Pirates

The meeting was again at the lawyer's office, with Jimmy, Alex, Alex's girlfriend, and myself. Our attorney, Walter Rose, was also present along with Sonny Cockrell representing the state, and one other state official. They took the plate out of the safe deposit box, placing it on the table. Sonny started talking. It seemed the state didn't have any money to write us a check. The reality was the state was in a delicate position. What they wished to do was to take the tray, period. However, they knew that would be a lousy move even for the state's low standards. They had no money to offer, and to secure the funds would take an act of legislation. We could have avoided all of this and still maintained the provenance of the tray by saying, "We found it on the beach." It seems so simple now, looking back on it. But we always thought honesty was the best policy. Fools we were. We were talking about real treasure, the largest gold artifact ever found on a shipwreck in Florida waters! No one had ever seen anything like that gold tray before the day we rescued it from its watery grave. Lives had been bet and lost securing these kinds of riches over the years. Why would it now be different? Greed still exists regardless of motive. No matter if it is in the quest for power or wealth, pirates still exist, they just have degrees and government titles. They might smell a little better in modern times, but their tactics still stink, and their hearts are still selfish and dark.

Well, this wasn't good news. But Sonny had a quick solution. He said he could convince the powers-that-be to trade gold coins in their collection for the tray. The amount of the trade would an equivalence of $25,000, the same amount I had appraised the value of the tray at during our first meeting. Wow, was I spot on or what! I did not really comprehend that my appraisal was so close to the actual value of the tray! I should have gone into the antiquities appraisal business much earlier, seeing how I had such a knack for it. The truth was that none of us knew our ass from a hole in the ground except for Sonny Cockrell and the state. Our attorney was no help. He challenged nothing, he researched nothing. He was happy to be along for the ride. There was only one document produced for the whole affair and the state produced that for him. None of the three of us who found the tray had two dimes to rub together. We worked from job to job, doing what we could to get down the road. We needed the money, we had little choice, we were young, poor, and dumb. Sonny opened his briefcase and produced an assortment of gold eight escudos recovered from the 1715 fleet wrecks, the same wrecks that had carried the tray as cargo. Some coins bore a full date, some a partial date, and some no date at all[1]. The date imprints have a major impact on the value of any Spanish coin manufactured in that period, prior to the production of New World machine pressed coins, circa 1730. The fully dated coins were beautiful and valued at $5000 each while the partially dated and undated coins were valued at much less. Alex, Jimmy, and the attorney all chose perfect, fully dated coins for their share at $5000 apiece. For my choice, I picked an eight escudo with a partial date of 1713 with just the "713" visible valued at $3000, and an undated eight escudo coin for $2000, summing my total share at $5000. The reason I did this was that I intended to sell one and keep the other. We took a few pictures of the deal, the coins, and the tray. We asked if Sonny had learned anything about the plate since we last saw him. He said not much, except it might be what they call a "glove tray". We exchanged a few more pleasantries, and the deal was done.

I would imagine Sonny was ecstatic all the way back to Tallahassee. He should have been. He just hoodwinked us better than any other conman might have. We asked him if he had any idea who might buy our coins from us. He could recommend one man in particular, by the name of Frank Allen. Frank had been involved with a group of other salvors in a successful Colored Beach 1715 recovery project and had become quite the numismatist specializing in early

---

1   Coins manufactured in Spain's New World mints prior to 1730 were struck by hand. No two coins are exactly the same, and the chance of finding a full date on any of them is rare.

Left to right: unidentified state official, Jimmy Ryan, Alex Kuze, Randy Lathrop, Sonny Cockrell, and Attorney Walter Rose. Photo: Susan Fletcher

colonial coinage[2]. Sonny also provided the name of another man out of New York City who dealt in early colonial coinage. We shook hands, took our reward, and walked out the door, leaving behind us one happy attorney who had a gold eight escudo in his pocket for doing nothing and two happy bureaucrats who had just secured Florida's most premier and prestigious artifact for pennies on the dollar.

In the weeks that followed, I called the man in New York who said he would pass through my area soon on vacation and would call me to meet up. We met at the Holiday Inn in Cocoa Beach a short time later. As I recall the gentleman's name was F. S.Werner and he was a Jewish businessman from New York City. He asked what I wanted for it and when I told him, he said, "No problem," glanced at the coin, and gave me my cash. We thanked one another and that was that. He had a very nice 1713 partially dated eight escudo, and I had my $3000 cash.

2 Frank Allen was an associate of Bruce Ward and Don Nieman. They were instrumental in the initial discovery of the Douglass Beach Wreck, south of Fort Pierce Inlet, which is one of the ill-fated 1715 Fleet ships.

The next day I drove to the Kellyco metal detector store in Maitland. At this time in 1978, Kellyco was a small shack near the railroad tracks. I first met the driving force behind Kellyco, Stu Auerbach, that day. I remember Stu showing me how to run the machine I bought. He used driving a car as an example... turn it on, this is your clutch, your gas, etc. The man who would become king of metal detectors taught me how to use my first metal detector. It was a White's detector and I've never bought a detector from any other man since. Over the years we also would become friends as I found that Stu was always nice, informative, and honest. What more could you ask for? I couldn't wait to get that detector working just for me alone. I was so happy about getting that machine, I just had to tell Alex and Jimmy:

"Why do you think you can find treasure on the beach when we've found very little with Chuck considering he has spent so much money looking for it himself?" Jimmy asked.

"Well, it was a beachcomber that provided Chuck all his information," I replied.

Well, "good luck" was Jimmy's attitude. Alex was sort of neutral on the issue.

I hunted the parks, numerous houses, and around old neighborhoods. My collection of Mercury dimes flourished, along with a few silver quarters. I was having fun and getting to know my machine like it was a part of me. A few months after our trade with Sonny, around Christmas, we heard that Sonny Cockrell was doing a presentation about the treasure from the 1715 Fleet disaster at Brevard Community College. The three of us agreed to go, and we met up at the lecture hall on the BCC campus. The event was well attended. Sonny took the stage and he had a slide projector. The lights dimmed and a large picture of the gold tray appeared on the screen. The audience gasped. We crapped our pants, as Sonny started his presentation. Sonny seemed to have gained a wealth of information on the tray in just a few months since we last saw him. We smelled a rat as Sonny continued with his lecture while the gold tray took center stage. The more he talked about what a wonderful once-in-a-lifetime find this was, the more we felt our stomachs churn. It was becoming vividly clear with each passing minute, and the newfound clarity made us question our actions and decision to do what we thought was "the right thing". We got screwed. He omitted how it was found or who found it, simply stating, "A diver found it." It looked like we'd all have to change our name to "A Diver" to get any credit, even a sliver. As the lights came up, Sonny suddenly saw us in the audience and we made eye contact. It wasn't good. He shifted his weight from one foot to the other while we stared back at him with our death-ray eyes.

"Oh, and the three boys who found the tray are sitting here in front... please stand."

He was forced into acknowledging us. The audience clapped politely, Sonny thanked the audience, then he immediately disappeared. We walked out feeling like we had just been slapped in the face. The state, to this very day, gives no credit to those who found their most highly-valued, premier artifact in their museum. When I last spoke to KC Smith, who was the docent or associate curator of the Gray museum in Tallahassee in 2017, she had a different story. She had been telling visitors over the years that a father and his two sons found the golden tray! I realized on my visit to the museum in 2017 that the powers-that-be are still smug, self-righteous, and inept. Nothing has changed, only festered.

The "platter/tray" on display at the Gray Bldg. in Tallahassee today. There is no attribution for the location where it was discovered, when it was discovered, or who recovered it. Photo: Alex Kuze

Chapter 7
The Dog Days Of Summer

It was 1979 and I was working as an animal control officer and cruelty investigator for Brevard County. Those days you never knew what you might encounter in the county. Rattlesnakes, alligators, monkeys, panthers, dogfighting... the job differed from your normal 9 to 5. The county supplied me with a truck and cages, radio, catch poles, mace, and tranquilizer gun. I was told to only respond to calls from the public and not to go around looking for violations on my own. Go out in the country, stay out of trouble, don't start being Barney Fife, we'll call you when we need you. I drove just about every road in the central part of Brevard County; I knew most of them. Driving a well-marked county truck provided me access to just about anywhere I wished to go. Even the beach was no problem. The other side of the coin was that I saw cruelty beyond human imagination... it was definitely beyond my own. I saw dogs with coat hangers grown into their necks, leather collars, rubber bands, wire, all left on the animal until it was grown over with oozing flesh. I've seen animals abandoned, left to die of starvation, forced to cannibalize one another. I've seen animals tied up, never attended to, left out in the rain, cold, and heat. One dog I had a call on was so covered in flies I couldn't tell what color the dog was. I wanted to strangle the woman who was responsible, My only recourse at the time was to ask her how she could go out in public and hold her head up knowing she had this wretched animal tied up at home. I dealt with stupid people and dangerous dogs. Many times the ones who suffered most were children, mauled by their parent's dog. I saw a family

Doberman take half a child's face-off. I witnessed a scene where a 3-year-old child had crawled into the family's pit bull pen in Cocoa across from the Jumping Flea Market where the dogs had killed the child. I can't pass that house even now without thinking about that day, and the scars it left behind on so many people. That little girl would have been a young woman now, probably with children of her own. We'll never know. That death was so stupid and preventable.

I was so disturbed by what I was seeing and very little being done about it. I went to State's Attorney Norman Wolflinger's office and asked if someone would show me how to file charges, instead of just confiscating the animal, which is what the normal procedure had been. I must commend the state attorney. It surprised him that the Animal Control Department didn't know how to file charges, and weren't doing it all along. He called the entire county Animal Control Department to a meeting one evening, and he went through the proper procedures to investigate and prosecute. I wasn't very popular with my superiors after that. Now they had to step up, instead of simply confiscating the abused animal and putting it down.

Brevard county is full of Diamond Back and Pigmy rattlesnakes. I became proficient with a catch-pole, that, I can assure you! People would often walk by and look into the caged area of my truck while I was at a convenience store or parked anywhere else to see what was in there. I loved it when I had a rattlesnake in a cage in the back and I would open the door and say, "Check it out." They would peer into the darkness, and that rattlesnake would let go with a good rattle. They would jump back with a start! We dealt with many exotics ranging from monkeys to tigers and that's the truth.

Being out in the fresh air, dealing with all the critters was fine by me but people were the problem, and I tried to avoid them. Not all people, some were a joy to know, like the Jordans on north Merritt Island. Drew and Ida Jordan lived just off SR 3 on Judson Road. You knew where to turn when you saw the hand-painted sign on SR3 that said, "wallamelons for sale". Another Animal Control Officer, named Bob Steimel had introduced me to Drew and Ida when I first started working for the county. They had property on the edge of a grove and had been there a very long time. Maybe they owned or inherited

The author, operating as an Animal Control Officer for Brevard County's Animal Control Division

the property or were just allowed to live there as caretakers, I'm not sure. I can tell you they were pioneers, and most of what they had or consumed came from the land. Drew had several old refrigerators that he converted to smokers and was always smoking mullet which he sold, along with melons and fruit. I would drop by and give them an opossum occasionally, which they really enjoyed. I watched Ida prepare them one day. She explained every step of the process to me. She would burn the hair off them first in a fire, then place them in boiling water. Ida said, "Honey, you get yourself a pot full of boiling water, some bleach, a wire brush, and you scrub them real good."

She went on with the rest of the process but she scared me off at the wire brush and bleach part. Drew did not know how old he was. I would guess in his eighties. I always stopped by and checked on them while I worked for the county. Merritt Island has gone through so many changes just like the rest of the state. I don't know what became of Drew or his soft-spoken wife Ida, but I'll always remember that "wallamelons for sale" sign.

The county required each Animal Control officer to be on 24-hour emergency call one week every four or five weeks. This week it was my turn, that's why my phone was ringing at 1 o'clock in the morning. "Hello," I answered. It was civil defense who acted as our dispatcher after hours.

"Cocoa Beach PD is asking for help with an alligator behind the Booby Trap Lounge," the voice on the other end said.

"I'll be right there," I told dispatch. I dressed and climbed in the county truck.

"235 is 10-8, 10-51 to Cocoa Beach." My call number was 235, and I had just told them I was in service and headed to Cocoa Beach. As I was driving I was thinking, this is weird, what is an alligator doing at the Booby Trap Lounge? The Booby Trap Lounge was an icon on the beach, a topless bar, that had a roof shaped like two big boobies. I pulled into the parking lot and was making my way around the south side of the building when I encountered a Cocoa Beach police officer walking to his car. "What are you going to do with that," he asked? He was referring to my catchpole I had in my hand. I was in no mood.

"Do you have an alligator here or not," I responded.

"You better come look at him first," the cop replied.

I followed the cop back behind the building, which was right next to a hotel. There were more cops standing around looking at something. They parted and moved aside when they saw me. When they moved, they revealed a huge gator. The alligator was every bit of 14 feet. The cops shared how the call started with me. A young boy who was staying at the hotel was riding his bike down the sidewalk behind the lounge and hotel. It was dark, and he hit something and rolled over it. When he got to his feet, he saw it was a huge gator he had hit with his bike. He told his father who didn't believe the boy but followed him outside to confirm, which he did and called the cops. I figured what had happened was that the gator had come from the Banana River through the Canaveral Locks, then through the port, and out into the ocean. It had swum south down the beach, gotten tired, and crawled up on the beach, probably looking for "sweet water", which the Crackers call freshwater.

I was going to need help and a bigger truck for this gator. I called my boss, Officer 231, to come help me. He was from New York, not much help but another set of eyes and hands, anyway. When Officer 231 arrived, we got to work. There was one of those long pool cleaning poles close by. I made a lasso out of some rope I carried with me, then I had 231 hit the gator on top of the head, just enough to annoy him. When the gator got hit on the head, he raised his head and hissed, that's when I threw the lasso over his head. I took the rope and took a couple turns around a close-by telephone pole. We then pulled on the rope. With the gator feeling himself being dragged, he started to roll, which is what we knew he would do. As the gator rolled and fought back, we fed him rope until he wrapped himself up. We tugged and tugged on him until he was exhausted and exhaled a very long hiss and a sigh. Once an alligator does this, they are worn out, and you have maybe five minutes before they get their breath again. Since I was a sailor and knew my knots, I got to be the one to tie him up. I got on the gator's back, pulled his front legs back and his hind legs forward, wrapped them together, then tied his mouth shut with the tail-end of the rope. If this is done correctly when you get ready to turn him loose, you pull on one end of the rope and it all comes loose at the same time.

A crowd had now gathered around the scene since the bar had just closed down for the night. I needed a knife to cut the rope and asked the crowd if anyone had a knife? I heard two or three switchblades open and were offered to me. Officer 231 had brought an open bed truck with a set of rollers and a winch to haul the gator into the bed of the truck. He went to get the truck, while I watched the gator and the crowd. The bar patrons were now clear of the bar and had gathered in a large circle around the gator. They started throwing bottles and cans at the gator which just bounced off of him. We were just barely able to get all the gator in the truck, we had to curl his tail in to get him to fit. It was now almost four in the morning as we headed our separate ways. Officer 231 was taking the gator to the shelter in West Cocoa to wait for Florida Fish and Game to take the animal off our hands, and I headed home. After a few miles,

I hear the radio come back to life, "231 to 235."

"Go ahead, 231," I said.

"This damn gator has come back alive and is swinging its tail rocking the truck all over the road."

"Good luck 231, I'll see you Monday," was my last reply.

****

I continued to dive on the weekends, mostly cave diving in north Florida where there was rarely a weather concern, so normally it was a sure thing. Although on occasions of heavy rainfall it would affect the springs, sometimes in our favor. When the floodwater above the spring attained a certain height where the hydrostatic pressure was equal to or greater than the flow of the spring, it slowed the outflow down, making it much easier to penetrate the cave as the current against you was reduced. If the hydro-static pressure increased beyond a certain point, the flow would reverse!

I still love north Florida, the homey country feel of the area, and the diversity of the countryside. Many of my old hangouts are long gone now. Steamboat Bill's restaurant, named for its owner — Steamboat Bill Smith is one such. My friend Chris bought the restaurant later from Steamboat Bill and ran it for several years. The place was popular with cave and spring divers; the walls were covered with large prints of underwater scenes of the springs, fossils, and artifacts found locally. It was a shrine to the great diving in the area. The place served typical southern fare; fried chicken, pork chops, fried okra, and so forth. Chris was also a dedicated fossil hunter. I wrote a story for *Treasure* magazine about Chris and his finds, and it made the cover story. I often dove with Chris and several of the local hunters in the Suwanee and Santa Fe Rivers hunting for arrowheads. Diving black water is like diving in weak coffee or tea. The water has a very brown hue with visibility anywhere from four to ten feet. The rivers hold many mysteries in that black water.

The NACD (National Association of Cave Diving) meetings were frequently in Branford, where, incidentally, Steamboat Bill's was located. Several regulars at the meetings would become world-

Chris Lewis shows off some of his arrowhead and spear point collection, gathered from springs and rivers in northern Florida.

famous. NACD member Sheck Exley for one. Exley was the first in the world to log over 1000 cave dives at age 23. In 29 years of cave diving, he made over 4000 dives. NACD member Wes Skiles is another. Beneath The Sea, Inc. is a nonprofit, volunteer, membership organization, recognized nationally and internationally as a source of education about the ocean environment, and they recognized Skiles as their "Diver of the Year" for education in 1996. In 2004, they awarded Skiles the National Academy of Television Arts and Sciences Suncoast Regional Emmy Award for his work on the *Water's Journey* series. I remember both men speaking to the members at the meetings. Exley died, aged 45, on April 6, 1994, while attempting to descend to a depth of over 1,000 feet in a freshwater cenote, or sinkhole, called Zacatón in the state of Tamaulipas, Mexico. I made several lasting alliances and a few good friends at those meetings.

I was logging five to six cave dives just about every weekend. Some of my favorite cave dives in north Florida were Little River Spring, Peacock Slough system, and Madison Blue. Madison Blue

The author is seen here in the foreground, followed by Mark McCloskey in the Little River Spring cave system along the Suwannee River.

was near the Florida-Georgia border, ten miles east of Madison on the west bank of the Withlacoochee River. We almost always had this place to ourselves, rarely encountering other divers. The spring bubbles up along the west bank of the river, just before a bridge, close to the highway. The spring basin was maybe 30 feet deep. There were two entrances, which took you into a cavern with a large breakdown, and a sandhill in the middle. Over the breakdown, on one side was a small entrance into the system. A hundred-foot back ran the main-line. The passage is of medium size with a low ceiling that rolls up and down, side to side, then branching off. Further down the line, it opens up into the Banana Room, then a large passage called the Monkey Room, a large cavern, then further back it leads to another large cavern called the Gorilla Room. Depths here can reach 150 feet in places. It was here I first met and became friends with Parker Turner, cave diver and explorer extraordinaire. I was surfacing from a dive in the system at Madison, and when I climbed up on the bank and started dropping gear, I saw this guy leaning on his truck. He was thin, and had wavy blonde hair, smoking a Marlboro cigarette. He looked me over real good, looked at my gear, checked me out, up and down.

"How was the dive?" he asked in a panhandle southern drawl.

I said, "Good," and noticed his truck was full of dive gear. "I'm Parker Turner," he said.

"Nice to meet you Parker. I'm Randy."

"You look familiar, have I seen you at NACD meetings?"

"Yes Parker, NACD number 215." This brought a big grin to Parker's face as Parker was the National Association of Cave Diving's safety director and was very active in the organization. Parker was the brains behind the Grim Reaper signs posted at the entrances to underwater caves showing a figure of the Grim Reaper stating, "DANGER go no further unless trained cave diver! You will die."

If you were at a spring cave, and it didn't look like you knew what you were doing, and Parker rolled up on you, it would sound like this. Parker always being the gentlemen would say, "Hi, hey listen, looks like your not cave trained, but if you insist on going on this dive, give me your car keys so I can give them to your survivors, save them lots of trouble, OK?" Parker was never shy about trying to prevent an accident in the making, he prevented countless deaths. Much more on Parker to come.

Parker Turner

Chapter 8
Bob Alone On The Dune

Chuck Kenworthy and his Quest Exploration crew only worked the summer of 1978 and we never heard much more from him or Quest Exploration Corporation after that. Unfortunately, the news came a few years later that our captain on the *Quo Vadis*, Captain Doug Dykes, was killed in a home-made ultra-light glider accident. Flying was another of Captain Doug's passions. I can still him waving that white hanky directing us where to drop the anchors. God speed Captain Doug.

On August 25, 1979, the National Hurricane Center reported that a tropical depression had developed within an area of disturbed weather, which was about 870 miles to the southeast of the Cape Verde Islands. It was that time of year again, the hurricane season. I'd been using my new detector here and there, in a park, around old houses, etc. The oldest coin I'd picked up was a 1909 Indian head cent I found when I crawled under a porch to detect on Indian River Drive in Cocoa. I was ready for my first storm coins, and I didn't need to wait long.

After crossing the Windward Passage, the storm, now known as David, struck eastern Cuba as a minimal hurricane on September 1st. It weakened to a tropical storm over land but quickly re-strengthened as it again reached open waters. David turned to the northwest and re-intensified to a Category 2 hurricane while over The Bahamas, where it caused heavy damage. Despite initial forecasts of a projected landfall in Miami, Florida, the hurricane turned to the north-northwest just before landfall to strike near West Palm Beach on September 3rd. The storm's broad center first touched land at midday September 3rd near Palm Beach, about 160 miles south of Cocoa, Florida, then straddled the coast throughout the afternoon as it moved north toward John F. Kennedy Space Center. Winds of 75 to 90 miles per hour struck the Melbourne area, 40 miles south of Cocoa destroying the beaches with erosion. The storm would arrive along the beaches of the Treasure Coast in the dark, with the storm center arriving over Cape Canaveral at midnight. I had been on the phone with Alex in Fort Pierce, who had been fortunate enough to receive a brand new White's metal detector for Christmas. We planned to meet at the sand trails leading down to one of the 1715 sites the following day.

Hurricane David left a path of felled trees, downed power lines, and heavily damaged buildings, bridges, and cars along Florida's Atlantic coast as it continued rumbling northward. A shift in the hurricane's path, which killed over 650 people in the Caribbean, spared Miami and the Florida Keys the deadly onslaught they had feared, tracking along Florida's beaches further north. The hardest-hit area appeared to be South Melbourne Beach, in Brevard County. There, a tornado associated with the hurricane ripped the roof and rear wall from the 18-unit Opus 21 condominium, sucking furniture and appliances from the building. About half the trailers at the Ocean Holiday Travel Park were destroyed, and the Melbourne Beach Police Chief, Evel Roberts, estimated damage in the hundreds of thousands of dollars. I was sheltering at my girlfriend's mother's house near west Cocoa, and checking on my mobile home periodically during the storm. All afternoon, the winds increased with a steady, ominous drone that rose to a high-pitched whistle. A visit to check on my mobile home and collect my metal detector was unsettling as the trailer ever so slightly come up and down an inch or two during a powerful gust. I didn't stay long at my place and driving in the winds was no fun with gusts blowing you from side to side.

Fallen trees and power lines littered Route A1A, and about 158,000 people were reported without electricity in Dade, Broward, Palm Beach, St. Lucie, and Martin Counties. With the winds and the damage came telephone reports from beach residents and some businesses that looters were moving in. The Sheriff's Office could not immediately assess how severe that problem was or would become. There were unconfirmed reports of looting in Satellite Beach, and the police there were reportedly arresting anyone without emergency

The track of Hurricane David along Florida's Treasure Coast in 1979

credentials. This made it difficult to get to where I needed to be after the storm. In the many years that have followed, it has become more and more difficult to get to the beach sites after the storm. (Near Johns Island many years later the cops were chasing away anyone but themselves from the beach and they got busted keeping the loot for themselves.) As it plowed up the coast, David seemed to move in fits and starts. After first brushing land at Singer Island just north of Palm Beach around noon, it advanced rapidly for about 95 miles to the Melbourne area. Civil Defense reported the eye of the storm over West Melbourne and Palm Bay at 9 in the evening. There, stalled by its contact with the mainland, the eye of the hurricane — an area of calm at the storm center — paused long enough to lure residents into the streets to inspect the damage. But by 10 pm, the winds rose again, and David resumed a slower advance up the coast, still packing winds in the 80-mile-per-hour range.

Ahead of the storm, the sea was surging into the first floors of condominiums near Cape Canaveral, and they called emergency crews to the Kennedy Space Center to tie down equipment and wrap the missile gantries in aluminum panels. By morning the storm had passed the coast. I woke up as soon as I could hear the winds backing off. Fallen trees, power lines, and other debris littered the roads as I made my way down south. I don't recall having any issues with the authorities getting across to the barrier islands; it depended on your route I suppose.

The sky had few clouds, and it was going to be a hot day like it seems to be after every hurricane with sweltering heat and dead calm. I pulled off the road to the entrance of the sand trail; I wasn't the only one with thoughts of treasure. In short order, I found Alex was there, and about maybe half a dozen others. I recognized John Brandon and maybe one more among the group. He and a few others were chatting with this person on the dune. Alex and I grabbed our machines and headed to the beach, walking by the group who were still chatting.

One of them looked over at us with our brand new machines, and said, "You boys going to find a little treasure today?"

The group looked over and chuckled, while we just kept our heads down and made way for the beach. I soon recognized the fellow who made that remark; it was Bob Marx[1] ! He was holding court, I suppose and he was the only one without a metal detector. We almost ran for the beach. When the others saw us going for it with no reservations, they ended the chat. We left Bob standing there, the only one without a machine! The others were now directly behind us as we jumped off the 10-foot escarpment to the black sandy beach

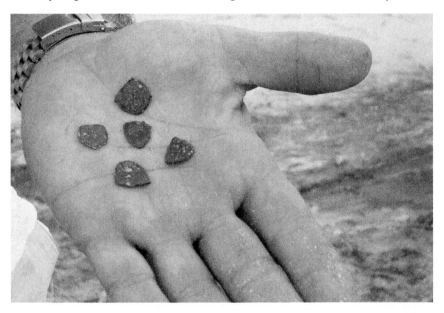

The author displays some of the 1 reale cobs he found. Note the black sand in the background, a sure sign that overburden has been scraped from the beach where coins are exposed.

below, wiped clean of debris except for the roots hanging out of the dunes. I didn't walk 50 feet and had my first signal. I dug frantically with my bare hands through the few inches of hard dark black sand, then my fingers felt the coin. I looked at it, after rubbing the sand from it. The cob was a 1691 one reale, Lima mint. I feared losing that small coin in my pocket so I popped it in my mouth. I knew if I kept

---

1   Robert Marx was a world-famous treasure hunter who wrote dozens of books on the subject and he was the operations manager of the Real Eight Co. for several years. He was well-known to many operators within the 1715 Fleet salvage community.

my mouth shut I'd be okay. I'm still learning that lesson, over and over, it seems. I soon found another half reale, then more one reales, then more half reales. My mouth was now full of coins, the taste of success! But I had to find another place to store my booty, so I finally trusted my pockets. I glanced up at the dune behind me, the others were now on the beach too, leaving Bob standing there by himself.

We picked up coins most of the day, with the majority of them being small denominations, half reales, 1 reales, and a few 2 reales. Lots of half reales were scattered along the base of the dune. I was working in "all-metal mode" on my White's because I didn't want to miss a single coin.

The once trackless beach was now covered with footprints and holes thanks to the small crowd of treasure hunters. In those days there were a half dozen or so "coin shooters" who would work a beach over, keeping a safe, respectful distance from one another.

The shoreline shown here is in the vicinity of Vero Beach after a storm. Uprooted palmettos and pieces of lumber litter the sand. Such storms produce optimal conditions for recovering treasures along Florida's Treasure Coast, or elsewhere, as the author would discover.

Nobody crowded the other fellow, and we kept our finds to ourselves. Unfortunately for the dedicated few who knew one another and kept quiet for years, things would change with time as they always do.

Site locations would be shared by print in books, and now online. A few folks would run home and call the newspapers, get their pictures taken with a handful of coins and gloat a little for their fifteen minutes of fame.

The local Johns Island police dept was busted for chasing beachcombers off the beach so they could have it all to themselves. Welcome to modern-day Florida.

An old proverb comes to my mind...

"The ability to speak several languages is a valuable asset, but the ability to keep your mouth shut in any language is priceless." – Anonymous. I Don't wish to sound like a hypocrite since I'm writing this book, at least I waited 40 years.

The sun crept higher, and it got hotter. I didn't want to leave the beach, and neither did anyone else, but the thirst and the heat were taking their toll on everyone. If you don't have a metal detector, just take a bunch of cold sodas to the beach! I would have happily paid a reale or two for a cold soda that day! The power was down, no stores were open, there was no water, no soda machines. It was brutal. The cursed heat finally drove us off the beach. I had done okay that day, I think I had close to 40 coins, all small, but treasure none the less. I don't recall what Alex did that day, and I'll just leave it at that. I suspect he needed to take that machine out of the box and get to know it sooner than he did. I drove home and was lucky enough to find a small store open with warm sodas and warm beer, cash only, and I had treasure in my pocket. Life was good, and I was ready for the next beach cut. I couldn't wait for the winter storms to commence!

Chapter 9
The Reale Originals And
The Money Path

I met Lou Ullian when I was trying to sell my second coin. Simultaneously I was soliciting offers for the coins that Jim and Alex owned. Lou introduced me to Rex Stocker, who also wanted to look at the coins as both he and Lou were very interested in anything relating to the 1715 Fleet shipwrecks[1]. Lou was the nicest man you could hope to meet anywhere, down to earth, unassuming, and a huge fan of miniature HO scale trains. He loved them. I have been to two of Lou's homes and he had a room with a permanent train set in both houses. Lou lived very close to Chuck Kenworthy's E56 site, and I'll bet he had helped Chuck out with a brief history of what rolls into that beach.

It thrilled me to meet and come to know Rex Stocker and his wife Carol. The nicest couple, once again genuine. I can't say enough good things about Rex and Carol Stocker. Rex has a cheerful, almost amused outlook on life and he seems more entertained by any issue or problem than annoyed by it. I had read all about Rex in the book "Pieces of Eight", and now we were on track to becoming friends. Rex lived over on the Sebastian River near Roseland. I docked my small boat at his house for a few weeks in the summer and I photographed his collection of rare treasure books for him. I still have copies of the slides of the beautiful colored illustrations.

Chuck's E56 lease encompassed an area very close to Lou's house, as I said, and very close to a path that leads through the jungle over the dune to the beach. The path was not well-traveled and

---

1   Rex Stocker and Lou Ullian were two of the original members of the Real Eight salvage team.

difficult to find; it got easier once you knew about the Australian Pines that marked the area to which the path leads. I remember my first time on the "Money Path". It was very much by accident. My good friend, Chip, introduced me to his cousin Lee Spence in November of 1979. Lee had heard about our gold tray find and was eager to meet me. Lee was quite the accomplished author and historian. He started salvaging confederate blockade runners, and writing about it before he was out of high school. Lee is an engaging character, soft-spoken with a very distinct Charleston accent, a descendant of a very old southern family from Sullivan's Island, South Carolina. Lee is well-read, especially about anything dealing with the civil war, or shipwrecks. Lee and I would work on several shipwreck projects together in different parts of the country.

Lee and I hit it off from the start. We had many things in common, with treasure and shipwrecks foremost, poverty and wanderlust second and third. It was November when Lee showed up at my house and asked me if I'd show him the area where we found the tray. I agreed to go on a dive with him and show him the general area of the imagined wreck site. Driving down A1A we found the weather was crisp and clear, and the seas flat calm as we pulled into my favorite sand trail. The water was icy and very clear, especially for November. The visibility was top to bottom and more in 20 feet of water that day, rare for this area anytime. We suited up in full wetsuits with SCUBA gear. We swam east from the beach accompanied by the sound of the popping, crackling reef, then turned north, then northwest, then south, following the same path that we had taken that fateful day on June 21, 1977. I was sure I was close to the tray area when we stopped. Lee had brought with him some hot pink survey tape, not sure why really, but he wrapped the nearby outcroppings with the survey tape. We wrapped the tape around three or four ledges in the area where I reckoned we found the gold tray. Then we swam south on the surface, looking down at the bottom as our air supply was fading fast and the winds were picking up from the north. As we swam, I spied a dead stingray on the bottom in about 20 feet of water; I dove to look at it as I was using my snorkel only now, as my SCUBA tanks were empty. Once on the bottom I looked over and saw what looked like the rim

The author is seen here with the Vanilla Bean Jug he recovered in the vicinity of the Corrigans wreck area the day it was found. Photo: Tom Kraft

of a clay jug, I swam over and gave it a tug, a tiny piece of the sand eroded rim broke off; I stopped, started fanning with my hand, then headed to the surface for another breath of air. Free-diving with full SCUBA on your back is difficult. The wind was increasing and I could barely see the object on the bottom from the surface at this point. The wind was trying to push me off my mark; I took a deep breath and dove once more; I fanned like crazy, gave it a gentle rocking motion, and the jug popped free from the bottom, intact! The size and condition of the intact jug from the 1715 fleet stoked me beyond belief!

The wind had picked up substantially as we kicked hard for the shoreline, with my jug held tightly against me. Once ashore Lee and I admired the jug which would be identified later as a "Vanilla Bean Jug". It was intact, with a small amount of sand erosion near the lip where I had broken off a small piece, but amazing none the less. To my knowledge, it is the only intact specimen of its kind to be recovered from the 1715 Fleet. I impressed Lee and I think he saw I had some good fortune or rather dumb luck hanging over me.

Lee spent the night on my couch, as he would many times. We hoped to get up early the next day for another dive while the

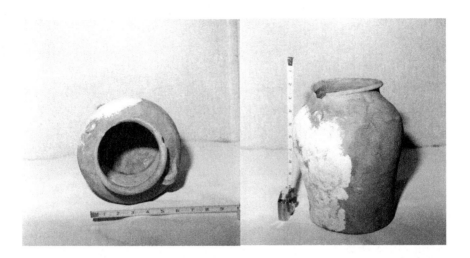

Details of the Vanilla Bean Jug, a one-of-a-kind artifact recovered by the author in the same area where he, Alex Kuze, and Jimmy Ryan found a gold tray. See Chapter 3.

water was clear. We drove south on A1A as we had the day before. While crossing over the bridge at Sebastian Inlet we could see the whitecaps on the water to our south. We pulled into the little sand trail, making sure not to get stuck; we pulled many a tourist from the sand on those trails. As soon as we crossed the dune and looked out at the ocean, we both knew it would not be a day for diving. The wind continued to increase as we stood on the beach looking out at the waves. We sat on the dune and ate the sandwiches we had intended to eat for lunch. Returning to the car we headed back north up A1A once more and as we crossed the bridge this time the wind had picked up and was blowing at a good clip out of the north.

As we drove along Lee was staring out the window. Suddenly he yelled, "Stop, stop the car!" He startled me.

I said, "What, what?"

He replied, "That's John's car. He's a friend of mine!"

I pulled the car over to the side and we got out, crossing over the dune to the beach. Just down the beach ahead of us was a tall figure sweeping the beach with a military mine detector. As we approached, I saw the figure look up and his face drop when he recognized Lee. "Hi John, how are you," Lee asked.

"Hi Lee, just killing time," was the man's reply, now identified as John.

"Having any luck?"

Before John could answer, I swear to God, I looked down and there was a green piece of eight on the sand below us at John's feet.

"Holy shit! Look," I exclaimed. John was not happy at this little unexpected revelation.

"Lee, please just keep your mouth shut, OK," John asked.

The tide was getting high, so high it chased us off the beach with the increasing waves. This was my first introduction to the "Money Path". This was in the same area as Chuck's lease E56. I didn't know it then but I had just met Chuck's beachcomber who had driven Chuck's passion for that area of coastline.

We drove back north to my house. "How many metal detectors do you have, Randy," Lee asked.

"Just the one Lee."

"I think my cousin has an old White's detector, I'll call him."

I could tell Lee was excited about the next coming high tide early the following morning. Lee's cousin had a detector, so Lee jumped in his car to go get it from his cousin in Titusville. Lee returned a while later with an ancient White's Goldmaster. It was old, but it worked. Lee didn't wish to place all his trust in that old machine, so he proposed that it was one for all, and all for one. We would pool our finds and split them down the middle. I agreed, feeling it was only fair... this time. We prepared our machines for the saltwater and the weather using turkey basting bags and duct tape.

It was still dark as I closed the car door. The wind was howling, and the temperature had dropped by 25 degrees. John was there again today, he was cordial but subdued, I wouldn't have blamed him if he hadn't had been such, as he wasn't thrilled to see us. We stood on the dune and looked down. The waves were full of debris. Pieces of wood and enormous tree trunks rolling in the surf, ready to take out any careless person who got in their way. The salt spray stung our eyes and covered our faces with a thin salty crust that tightened our skin. We stood there on the dune for some time before we mustered up the

The beach somewhere along the Treasure Coast is seen here after a hurricane. Note the black sand, which is the tell-tale indication that overburden has been moved away from the underlying clay where coins can be found.

courage to jump down in the cold splashing, violent surf below us.

The key, I soon learned, was to never take your eyes off the surf, and count the set waves! I hadn't been on the beach for over five minutes when my headphones about jumped off my ears with a loud, solid beep. I quickly reached down with my hand, probing the sand for the coin and looking out to sea for the next large set wave. My fingers closed around a heavy 8 reale. There is no other feeling like it! The problem was, and it was a big one, was, if you took your eyes off the surf you would soon become a victim to its fury, every set wave was looking to take you out and pull you under. The cold water would come rushing at you, engulfing you up to your waist, sometimes chest high. When you saw a set wave coming at you, you had to jump up high on the eroding dune, grab a handful of exposed roots and hold on for dear life. I had one signal after the other, soon I had quite a few coins, all 8 or 4 reale denominations. Lee would come over to me and ask, "How many you got?"

"I got about six. How many do you have?"

"I have about four," Lee said.

And so it went for the next several hours. We picked up coins until we were cold and convinced we had done all we could to get our share. I can't remember the exact count from that trip, but I know it was over fifty coins, all solid 8 and 4 reales. I know when we got back to my house we woke up my roommate, debating who got what once we started dividing the coins. We cleaned the coins to see what we had. About a month later I went to pull the chain to my kitchen exhaust fan to turn it on and the entire fan fell out because of sucking all the fumes from the muriatic acid we used that day to clean the coins. This was the first nor'easter after Hurricane David eroded the beach in September, so the chances of another similar storm cutting down the beach even further were good!

I didn't have to wait long, January rolled around, and with it, another nor'easter setting up along the coast. This time a freezing cold front turned the ocean into a frothing monster of white water chewing away at the beaches. I showed up by myself this time; I walked the little hidden trail through the dune scrub and saw four individuals standing on the top of the dune looking down at the raging waters. I soon recognized two of them, John Watkins was there, along with Lou Ullian whom I'd met once before when I was trying to sell Jim and Alex's coins for them. Lou recognized me and introduced me to the other two men standing beside him; "Randy, this is Rex Stocker," Lou said. Lou then turned to the other man and said, "This is Bill Saurwalt." We exchanged pleasantries and continued to stare down at the beach, watching the beach wash away with every incoming wave. I was checking out Bill and Rex. They both had military metal detectors like John used. Bill Sauerwalt reminded me of Bluto from the Popeye cartoons. Same beard, same look, but thinner. Rex seemed to smile all the time. He was humble and just emitted a good vibe.

The ocean swirled and crashed against the dune below us. We could see many types of flotsam, tree trunks, lumber from destroyed cross overs, and many sorts of debris floating in the foam-covered waters. The waves would come almost to the top of the dune and then recede into the ocean with equal speed. We chatted and waited for the tide to back off as the waves looked too dangerous to jump

The author shows off some cobs found on the beach after a hurricane. Photo: Tom Kraft

down fifteen feet to the beach below. Lou started telling me that Bill had a wreck up north and hunted the beaches near Daytona, and I knew who Rex was. I was thrilled to be standing next to all of them on that cold blustery morning. As we chatted someone yelled, "Look!" We all looked down just in time to see a green piece of eight become exposed as it was being sucked back into the ocean. As soon as it had appeared, another wave rolled over it and it disappeared back into the surf. That was all it took, high water and waves be damned. We all started jumping off the dune like lemmings, one right after the other.

I wasn't on the beach more than a few minutes when my headphones about blew off my head with a solid ring; I reached down in the wet sand and felt the one-ounce piece of silver just a few inches under the sand. This spot produced only eight and four reales, and they were solid coins, green and chunky. I pocketed the coin and took another swing, another signal with the same results! Trying to keep one

An 8 reale found on the beach

eye trained out to sea, watching for waves and the other on the beach looking down was tricky. You were in constant danger of being sucked out to sea by a large set wave if you didn't keep an eye out on the surf. When you spotted a gigantic wave coming at you all you could do was jump up and grab a root protruding from the dune and hold on for dear life. The water from the wave would sometimes follow you, rushing at you and coming waist-high to you, even hanging on to root halfway up the dune didn't spare you all the time. It didn't matter; with every wave that rolled in, a coin came with it. I could feel the coins hitting my feet and shins when a wave came through. This is no lie. I would have a hit and before I could pick it up a wave would come in and move it several feet down the beach from where it was before. I would find it again, and again before I could pick it up another wave would move it once more. I chased coins sometimes as much as 50 feet down the beach before I could get them in my pocket.

My pockets were getting full and the weight of the coins required that I tighten my belt another notch just to keep my pants up. I was forced to go back to the car and empty my pockets three times! No one else showed up that day, just the five of us, and it was nonstop for a solid two to three hours. We were cold, wet, and grinning from ear to ear when we walked off the beach down the path to our cars that day. My last count was 50 eight reales and 18 four reales. Sixty-eight pieces of Spanish silver. Not bad for a day's work! At a conservative estimate of a hundred dollars a coin, I just made $6800. I must mention that this was during the Hunt brother's silver buying spree, so silver was $60 an ounce. It was easy selling coins for a hundred or a hundred and fifty per coin. I quit my job. I now had the grubstake to look for "more engaging" employment.

Chapter 10
A Good Year

My friends Jimmy and Tom, fellow students of photography and dive buddies, were shooting pictures in Daytona for the race track[1]. When they returned I showed them my haul. Their eyes lit up, and Jimmy swallowed what he'd said about my efforts previously. I soon received a call from Jimmy asking me to accompany him over to Kellyco to get a new metal detector. Two weeks after that I received a call from Tom asking me the same thing. Everyone now had new White's metal detectors, and we were waiting on the next blow. The month was now February, and the meteorologists were calling for another nor'easter, right at the full moon. If you don't have the full moon, we've found you are wasting your time.

The storm was a strong nor'easter, stronger than the storm before in January. We bought new turkey browning bags to wrap our detectors to protect them from the saltwater. We found the browning bags worked well, they were the right size, and hard to tear. And duct tape, the lazy man's best friend. We wrapped and taped, and wrapped and taped some more until we were satisfied that we had outsmarted the elements. We loaded into my 64 Chevy and south we went, stopping by McDonalds for an Egg McMuffin per our tradition of good luck. We arrived at the chosen spot and ran down the narrow trail surrounded by sea grapes to the dune and the ocean. The seas were enormous and slamming against the dune, sending salt spray high into the air. It was cold, and it was raining. It was the cold that went right through you to your bones. Florida can get cold, cold enough

---

1 The world famous Daytona 500 racetrack

to kill a person from exposure. Just ask the survivors of the wreck of the *Reformation*, which sank in 1696. As they struggled north to St. Augustine in November, they lost several to exposure. We had prepared for the cold water by wearing our wetsuit "Farmer John" bottoms under our clothes[2]. Once again I stood at the top looking down into the maelstrom below us, trying to decide when to jump. I took the lead and jumped, with the others soon following my lead.

The waves and the undertow were even stronger than the last storm in January. I had my first signal in before I'd walked 20 feet, and they kept coming, one after another. The coins were there, the hard part was trying to pick them up. As soon as I would get a hit I would bend over to pick it up only to have an enormous wave sweep in and steal it back it away. This was exactly like my first experience here, the waves trying to sweep you away; the coins traveling down the beach with each successive wave, the feeling of coins hitting your legs with each huge wave. I looked around and it seemed like the others were busy with signals of their own. As the high tide increased, so did the size of the waves breaking on the beach. It was critical that you paid constant attention to the incoming swells and the enormous set of waves now crashing upon the beach. I had plenty of signals, but getting to them in time was the problem. More and more, I looked up just in time to see an enormous wave coming at me with barely enough time to scramble high on the dune and save myself from being washed over and sucked out to sea. My pockets were filling up with coins, and I couldn't feel the cold or the freezing water anymore. I was amazed! I was bent over after another coin when I looked up and saw a set of monster waves coming at us. I yelled at Jimmy who was bent over intently trying to locate a signal to warn him. He never saw it coming. I scrambled up the dune to safety where I watched the wave wash over Jimmy; he disappeared under a wall of water. As the wave receded, an arm with a metal detector held straight up above it appeared, much like the scene from the movie

---

2   A full-body wetsuit is usually in two pieces, and the jacket might be worn over the "Farmer John" portion, which is generally a single sleeveless piece that provides leg and torso coverage, hence the appearance of a farmer's bib overall.

Excalibur with the arm and the sword pointing skyward appearing from the depths. Jimmy then appeared on all fours, crawling towards the dune with his headphones dragging in the sand behind him. His machine was honking, beeping, squawking, and making noises never heard before as he crawled up the dune soaked to the bone. I continued hunting, feeling assured Jimmy was wet but OK.

The coins never stopped, neither did the waves, keeping it exciting, difficult, and dangerous. Everyone found coins that day, some more than others. Jimmy toweled off and motioned me over to the dune. "Hey, if you get tired, can I use your machine?"

I looked at Jimmy and replied, "Are you crazy? Go to the car and try holding your detector under the exhaust to dry it out."

Jimmy disappeared over the dune, headed to the car. I kept detecting, kept getting signals, continued chasing coins down the beach, back and forth, back and forth. It wasn't long before Jimmy reappeared on the dune, flashed a thumbs-up, and jumped back onto the beach and into the game again. We were wet and cold, our hands and feet were numb, but we hunted until we felt we had done all we could for that tide. We felt little remorse about leaving the beach under the circumstances.

The mystery that surrounds the money path and that beach is the lack of any shipwreck material beyond the coins found there. The eight and four reales in good shape were many, but nothing else. Not a nail, not a spike, not a musket ball, no pottery, no lead sheathing, nothing but coins. Why?

I have heard two theories. The first is that a portion of the forecastle bounced here, which I'm reluctant to believe. The second theory is that the governor of St. Augustine sent a longboat down to the wreck site to secure as much silver as he could as soon as he could. Most likely to ensure he would recover his cost and expenses for lending aid and supplies to the survivors at the very least. The story goes that a boat loaded with coins left the salvor's camp just south of modern-day Sebastian inlet en route to St. Augustine when the wind and seas picked up, driving them ashore and flipping the boat in the surf at the Money Path site. All the chests of silver were lost in the surf. This story supposedly has historical backing in the records and is the one I believe. The longboat flipping over in the surf would explain many things as to why there is no ballast, rigging, nails, pottery, musket balls, sheathing, etc. It would also explain the larger denominations of four and eight reale coins, with no smaller denomination coins. Imagine, who wants to get paid in dimes?

The coins I sold provided me the time I needed to find a new job. The new job was right in my backyard, working for a company called Aqua-Tech located on Cox Road in Cocoa. I had a lady friend whose father was the boss of this company, which lay underwater power and telephone cables in the sea bottom across inlets, harbors, and the Intracoastal Waterway. The company had built several of what they referred to as underwater plows through which they fed the power or telephone cable from a barge above down to the plow that would then saturate the submerged subsoil and insert the cable into the incision in the seabed. The company used two divers, and these two also took turns tending one another when not diving. So, all told, there were two divers on the job, taking turns tending one another, or working underwater. The divers would ride the plow as it was being pulled across the bottom, making sure there were no obstructions or problems with the cable being buried in the seafloor. The company used surface-supplied air to divers using Desco "Pot Hats", which were heavy spun copper helmets with one rectangular view plate in the front. The helmets were what we called "free-flow" helmets, meaning their air supply was unregulated and you could get all the air you wanted. They were a good hat for hard work. Being attached to a hose from the surface also meant that a phone wire could be served into the hose and safety line umbilical, providing a means of intercom communication to the diver from the surface tenders on deck. When deployed, the operation entailed several sectional barges, water pumps, two 22-foot work boats, compressors, and several semi-trucks to get us where we needed to go. It was an aquatic traveling circus.

I joined the Aqua-Tech crew, and found them to be quite a collection of characters. My boss George was a hardworking man, honest, patient, and willing to show you what he wanted to be done. George knew nothing but work, it's what he enjoyed, it's what he

Aqua-Tech crewmen George Clark, left, and Skipper Ware are seen here at an underwater trenching job at Indian Rocks Beach in Florida. Skipper is holding a DESCO "Pot Hat" as he prepares to dive. He is wearing a U.S. Navy Mark V weight belt, which is quite heavy. He is also dressed with a harness that is used to attach the dive hose to his torso so that he may be pulled to the surface in an emergency.

Diver Skipper Ware has attached his "Pot Hat" to its neckdam and George Clark is dressing him for the dive. Commercial dive helmets are equipped with microphone/ speakers that use hard-wire intercom communications. In the background you can see the diver intercom, in this case a Hellephone model, widely used in the industry.

wanted to do more than anything else. The story goes that George and his wife rented a lovely cabin in the Smoky mountains for a much-deserved retreat. They unpacked and while George's wife was making some ice tea; she heard a lawnmower startup. It was George... he had located a lawnmower in a shed and was happily mowing the yard. George relaxed by working, not much else brought him joy. George had a partner named Doug. They worked together at Bell Labs up north doing the same type of underwater work but on a much larger scale. Doug and his son Roddy both worked at Aqua-Tech, but Doug was just leaving to work for another company when I arrived. Roddy stayed on. Roddy had just turned 18, was full of energy, smart, witty, and just learning to drink (alcohol). Roddy was a deckhand/boat operator and sometimes worked as a tender.

And then there was Captain George Clark, straight out of a Carl Hiaasen novel. Captain Clark was a long time mariner, shrimp boat captain, teller of tall tales, romancer, and dreamer. Captain Clark was from the panhandle area of north Florida, picked up from

a job the company did in Carrabelle, Florida many years ago. George looked like Willie Nelson and possessed a similar raspy, nasally, slow southern cadence in speech. George's uniform was a huge floppy straw hat, with a generous slathering of zinc oxide covered his face and lips, making him look like a ghost. It protected every inch of George from the blazing overhead sun. George had a previous bout with skin cancer and wasn't taking any more chances. Captain Clark was hard on me at first until he figured out I was eager to learn anything he wanted to share. Once he warmed up to me, he taught me all the knots I ever wanted to know, and which ones you could bet your life on and which ones you wouldn't want to. George loved folklore and colorful history, and he loved Florida as much as I do. Capt George was the best friend we single guys could ever have hoped for and he was what we called a "starter". Whenever we would pull into a new town for a job, the first night we would fan out separately looking for the best watering hole with the prettiest waitresses. It was a contest we held every time we landed somewhere new. George was always faithful to his wife, but the girls loved him and his easy manner. George would tell them why he was in town and that he was working with a bunch of young buck divers that the girls should meet; George laid beautiful groundwork for us. George learned the

Above: the author in route to an Aqua-Tech jobsite

Below: deckhand Roddy Steen at the wheel of an Aqua-Tech tender

preferences of each co-worker where the ladies were concerned and after a while had us matched up before we even met the girls.

Skipper was the main diver when I came aboard. He was dark complected with sad eyes. Though small in stature, he was solid.

Skipper came from Virginia and had been working with Aqua-Tech for some time. He told a funny story about diving on a salvage job: he jumped overboard forgetting his flippers. One of the crew yelled out, "Hey what about your flippers?" Skipper yelled back that he didn't use them, but thanks anyway, trying to save face. So Skipper did the entire dive without flippers. He said it wore him out.

We had a new diver come aboard directly out of commercial dive school. There was not a scratch on any of his gear. The first thing we looked at when a new diver showed up on just about any job was his gear. If the gear looked worn, torn, like it had seen better days, it impressed us. A wet suit full of holes, patches, and duct tape told us all we needed to know. We knew we had a seasoned, experienced diver. If he had new gear we figured we had a rookie on our hands. This new diver served as tender until we knew him better, but he complained all the time about not diving. He went to dive school to be a diver, not tender. We had heard this too often and were growing tired of it. So we came up with a plan. Skipper was in the water and the new guy was tending when suddenly Skipper shrieked and said, "What the hell was that?"

The new guy who was tending Skipper asked, "What Skipper?"

Skipper replied that something large had just brushed up against him, but since there were only a few feet of visibility, he did not know what it could have been. A few more minutes passed and again the same thing happened. A short shriek from Skipper indicated he had some concern about "something" that was in the water with him. Now Skipper told the new guy to have me suit up and get ready to take over the dive which I did immediately. I went to work below with this new guy listening to my breathing over the diver com (intercom radio). After several minutes, I repeated the routine. "Holy shit, what the hell was that?"

"What, what?" the tender asked me.

"Something big just brushed up against me," I replied. After a few minutes, I repeated my same surprise and concern. I continued to work for several more minutes, muttering to myself. After a little more time, I asked the tender if he was ready to suit up and finish the dive. He was quick to reply that it was okay. He felt he could use a little more time tending before he dove. He never pestered Skipper or me about diving anymore.

I learned a lot from this crew, and we usually had fun. I have splendid memories and marvelous stories working for Aqua-Tech. One last story out of many I leave you with. On a job in Virginia, near the old naval yard in Norfolk, the company was laying cable when we learned that the Navy had kicked a headless corpse off the bottom when routinely starting up some mothballed ships to keep the props from seizing up. So the bottom line was to keep your eyes open down there for a head with no body attached. The visibility was just awful, maybe from a foot to two feet, just brown, green, nasty muck. We joked about who the lucky diver was going to be to find the head. This didn't amuse Skipper much as he was the one who spent the majority of the time in the water. So we worked along for several days and the subject faded away for most of us, but not Skipper. Tending can be one of the most boring jobs most of the time, listening to the constant breathing on the diver com, and letting out or pulling in wet hose. I was tending Skipper, just about to fall asleep when over the diver com came a hair raising, blood-curdling scream from below. I yelled into the diver com, "Skipper, Skipper, you OK?"

"I think I just found the head, standby," Skipper exclaimed. "I'm coming up, pull in my hose."

I started pulling him up. He slowly climbed the ladder with something in his hand and threw it on the deck. The object had long black hair falling down like the hair might on a shrunken head or one of those troll dolls some folks hang from their car's rear-view mirror. At first sight, it was a ghastly looking thing, but once you focused on the object it slowly became apparent that it was a paintbrush, an old waterlogged large paintbrush that the bristles had lost all their stiffness and now fell just like hair. It had floated within inches of Skipper's face and in the poor visibility, it looked just like a head floating by! All of us were so relieved it wasn't a head that we didn't laugh until talking about it later that evening. The job finished and the head never appeared, much to our relief. That paintbrush sure looked spooky, I can attest to that.

Chapter 11
East Coast Research

I had decided to sell my remaining gold coin, and Jim had asked me to get a price on his coin. I reached out to the man which the state archaeologist Sonny Cockrell had recommended. His name was Frank Allen. Frank was previously involved with the Colored Beach Wreck, in company with Bruce Ward, and a few other fellows. I'm not sure I understand exactly how, but I believe he raised some investor money for the project. He became an authority on early colonial numismatics, primarily Spanish coins, and collected, traded, and wrote about them. I spoke to Frank several times on the phone and we agreed to meet at a McDonalds in DeBary near where Frank lived. I liked Frank the very first time we met. Frank was a large man, with jet black hair combed straight back, large dark penetrating eyes, with heavy eyebrows, a friendly manner, and a deep, jovial voice. He had also been a schoolteacher and a salesperson during his life and had picked up useful traits from both trades. Frank always remembered your wife's or girlfriend's name and would always ask about them.

Jim had come along with me to meet Frank and get an appraisal on his coin. We settled into the booth at McDonalds, and to break the ice Frank told us to hold our hands out palm side up. He then dropped a nice fat gold eight escudo into both of our hands from a height of seven or eight inches. We would soon learn that this was Frank's calling card... "hold out your hand", Frank would then drop an ounce of Spanish gold into your open palm. We showed Frank the coins we had to sell and he examined each one carefully. Understanding that no two were alike, he took his time. Frank said that he was involved in the salvage of another 1715 wreck just north of Fort Pierce Inlet. The other individuals Frank was involved with on the site were not working out, and Frank hinted that there might be crew and diver opportunities soon. Frank and I agreed on a price for my coin, we also agreed to meet the following week when he would have the money for the coin. We shook hands and went our separate ways. As soon as the car doors shut Jim started in on me implying that I shouldn't have left my coin with Frank. Was I crazy? I countered that the state archaeologist had recommended the man as one of the few experts in Spanish coins, and I'm sure trust was part of being an expert! Jim wasn't going for it anyway and thought I'd lost my mind. The following week I drove back to DeBary and met with Frank once again to get my money. I liked Frank from the beginning, and he took to me. He and I became friends and associates, sharing plenty of time with one another.

I continued to work for Aqua-Tech diving in the most polluted, dark, nasty water you can imagine. Our job sites many times were rural remote areas with little to offer except small motels and

The 8 Escudo the author sold to Frank Allen was one he received in a trade deal with Florida State Archaeologist Sonny Cockrell. It bears the assayer mark of Jose Estaquio de Leon and the designator of the Mexico City mint. Most of the coins found on the 1715 Fleet wrecks bear this assayer's mark.

diners with friendly local waitresses. As fiber-optics took hold in the industry, our workload soared. We spent many days in the Florida Big Bend area around Apalachicola, East Point, Port St. Joe, and Panama City. I was making good money, but the diving sucked. I hadn't seen any decent visibility more than an arm's length in some time and was secretly contemplating and hoping for a chance to work with Frank and his company, East Coast Research. The boss and the crew at Aqua-Tech were the best I had ever worked with, and it would be tough for me to move on. Frank and I kept in touch on a weekly basis. He would stop by the house from time to time on one of his many round trips to Fort Pierce. Frank eventually told me that his other partners were ready to throw in the towel, and it looked like he would need an entire crew for the upcoming summer salvage season. I asked Jimmy, Alex, and my friend Tom Kraft if they were interested in joining the crew, and they agreed they were interested. I shared my thoughts with my boss George at Aqua-Tech. As always, George was a stand-up guy and offered me encouragement. He told me to always be the best I could be making it all the harder to walk away from such quality and integrity that he and the crew possessed. It was rare, as I would find out. I got the call from Frank notifying me of when and where he would need his new crew to be. I said my sad goodbyes to the Aqua-Tech crew to join another that I only hoped would be half as good as my old crew.

Frank had excitedly shown us pictures of the new boat which he had acquired from Perry Submarine in West Palm Beach. It looked great like a nice, big, spacious vessel. The day finally came and we gathered at Mel Borne's Marina at Fort Pierce Inlet to welcome the new boat to the dock. As we stood on the dock, Frank said, "There she is fellas."

We looked out. "Where Frank?"

Frank pointed at a strange-looking craft, maybe 29 or 30 feet long, open-top except for a small wheelhouse, with a hydraulic transom like we had never seen, making its way toward us. It was a strange vessel. Our hearts sank.

I finally found my voice. "But Frank, this doesn't look like the boat in the pictures you showed us."

Frank reached into his pocket and pulled out the small picture he had been showing us all along. "Sure it is, here it is fellas, just like in the picture."

We took the picture and looked again and, sure enough, there was the little boat, docked right alongside the big beautiful boat we thought was our salvage boat. My heart sank. Had I left an out-

The Ocean Runner was a small utility craft originally built to retrieve small submersibles. Photos: Tom Kraft

standing job for this? They called the boat the *Ocean Runner*. Perry Submarine built and used it to launch and retrieve its small submersibles. It was a funny looking thing; the transom would drop, allowing a small submersible to be drug on board by a winch. It came with its own skipper and mechanic. The skipper's name was John Kimbrell, a Florida native and employee of Perry on loan for the summer to Frank as part of the boat deal.

Our crew now included Jimmy, Alex, our friend Tom Kraft, John, and myself. Also, a magnetometer operator out of California joined us later on. Much to our regret, we went straight to Cracker Boy's boatyard to outfit the blowers and get the boat ready for the season. If you've ever spent time in a boatyard, you understand how awful it can be. The ground in a boatyard is full of toxins, paint, lacquer, solvents, fuel, etc. There is no shade unless you're hiding under the

Left to right: Randy Lathrop, Alex Kuze, Kirk Purvis, Jim Ryan, and Frank Allen
Photo: Tom Kraft

81

boat seeking relief from the tropical Florida sun. It's dirty and dusty, not where you want to be in the summer. The blowers were going to be a challenge to secure to this boat because of the hydraulic transom. Frank had rented us a small apartment close to the marina for a few of us and we settled in. We spent days installing the blower on the *Ocean Runner*, and finally, the big day arrived when we took it to the site and tried it out. It was a beautiful sunny day when we motored out the inlet on the *Ocean Runner's* maiden voyage. The anglers on the rocks at the inlet gazed out at the odd-looking vessel making her way out the inlet.

We arrived at the wreck site. We set the anchors to secure us in position, two fore, and two aft, to lock us in a four-point mooring, holding us where we wanted to be. We lowered the blower into place, locked it in, started the engine, and slowly throttled up. The clouds of sediment swirled below us, creating an underwater tornado sending bits of shell and sand to the surface. John increased the throttle, the

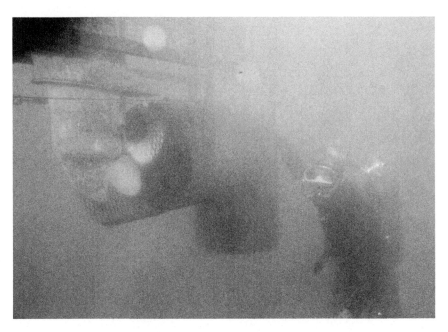

Shown here in its fixed position, the blower assembly on the Ocean Runner was unusual in design because the vessel used an outdrive for propulsion rather than a fixed propeller shaft. Photo: Jim Ryan

A large collapsible-stock anchor is seen here on the deck of the Ocean Runner. This anchor is not an artifact of the 1715 Fleet ships, being more common on sailing vessels of the 19th century. Alex Kuze is standing and to the left are Kirk Purvis, author Randy Lathrop, and Tom Kraft, right to left. Photo: Jim Ryan

boat shook a little, no big deal we thought. We increased the throttle a little more, the sand was boiling under us now, shell and sand swirled below us. We were digging full bore now... suddenly a big "Ka-Pow", the boat shook, and the blower shot right off the back and sunk to the bottom. We looked on in disbelief, happy that none of us had been working under the blower when it shot off the back. We suited up, hauled the blower to the surface, set it on the deck, and headed back to the boatyard, our heads down. As we passed by the anglers we had passed a few hours ago we looked the other way. Back to Cracker Boys boatyard.

The blower attachment went through several versions until we finally had one we could trust. The day we finally arrived on the wreck site again, the water was crystal clear. Anchoring up on the wreck site this time, we could see the sizable ballast pile below us. The ballast was concentrated, with some scatter here and there. Frank had

The Ocean Runner is seen here in the vicinity of the "Wedge Wreck" where there was an inlet at one time. Alex Kuze is standing on the boat's blower assembly. The Wedge Wreck had been worked over pretty thoroughly by Kip Wagner's Real Eight salvage group years before East Coast Research excavated the area. Today it is a designated underwater park and no further salvage is permitted. Photo: Jim Ryan

several areas to the southwest of the ballast pile he wanted investigated, so we concentrated on those areas first. Diving in the summer just off the beach from Fort Pierce southward you can expect clear warm waters, with plenty of marine life on the shallow reefs. I couldn't have been happier with my new diving environment. During the summer, it's common to encounter large numbers of big "Moon" jellyfish pulsating through the water near the surface. They are beautiful creatures, looking like large inverted pink bowls with frilly lace around the outside for trim. Normally they are of little consequence to swimmers unless you place them directly on top of sensitive body parts. However, we soon found out that when they float along the surface they get sucked into the props and get chopped up and all the little jellyfish pieces would get blown down on the divers. It was like getting hit with a hundred sweat bee stings all at once. The first several times it happened we would come up, strip off and wash off with fresh water. After a few days, it didn't matter anymore, we would laugh and help a big jellyfish into the props, messing with whoever was the unlucky one on the bottom. Another nasty little pest was the "bristle worm", a small caterpillar looking creature that had a super fine covering of clear bristles, barely visible when you had the misfortune of getting one in your hand. You had to hold your hand up to the light so you could see the superfine mono filament-like bristles in your fingers, it was impossible to get them out.

We loved Fort Pierce. The water was clear, living on the beach was fun, and the girls seemed to be happier to see you. We were getting paid $50.00 a day, and we only got paid when we went out to the site. We became friends with many of the charter captains, and they would save us the large grouper and snapper heads. We would boil or roast the heads and get plenty of meat from them, which we would add to tomato sauce and spaghetti. We were just as happy as if we had good sense. The crew had a fresh addition when Kirk Purvis from San Francisco showed up to do the remote sensing survey work with his custom-made cesium magnetometer. Kirk was a crusty old harbor hippy, living on a converted military boat in a harbor around the San Francisco area. Kirk had worked with the Platoro Group on the 1554 Spanish wrecks, a fleet of ships laden with treasures that sank near Padre Island in Texas. Frank had contracted a vessel from Walt Holzworth, whom he knew from working with Mel Fisher and his crew at the Colored Beach wreck. So now it was Jimmy, Kirk, Tom, and I at the apartment while Alex lived at his house in Fort Pierce, having moved there a couple of years before for construction work. John Kimbrell would drive back to Palm Beach most nights but stay

Kirk Purvis is seen here running a magnetometry survey. Note the paper chart plotter.

in Fort Pierce occasionally. Frank would run our pay down to us every two weeks and he would struggle many times to get us paid, but he always paid us on time. We split the crew assignments; Alex and Kirk on the mag boat and the rest of us on the *Ocean Runner*. We moved ballast from the main pile to either one side or another to examine what laid below and in those days the things to be found under the ballast were substantial. The local Sea Scouts, had removed a sizable amount of the ballast already, and they piled the stones at the marina next to us at the harbor.

Fort Pierce in the late '70s was humming with salvors in the summer. John Brandon and Art Hartman were just two that would pass us daily on our way out the inlet. Art Hartman had a crew boat painted white with the orange Coast Guard-like stripe running down the bow to discourage pirates in Caribbean waters. Art always ran north to the Corrigan's site, and John would run to Corrigan's or Colored Beach. On rare occasions, we would see Jeanne Durrand, who had a site just north of the Frank's Wedge Wreck site. Jeanne was a chain-smoking local legend who lived on her salvage vessel the *Barty Roberts*. Jeanne had done many things in her life, from parachute jumping to racing motorcycles. The *Barty Roberts* was an old shrimp boat that spent most of its time at the dock.

We spent that first summer getting familiar with one another, the operational procedures, and the wreck itself. Frank had us mag north and southwest of the wreck. Unfortunately, this area was a training ground for frogmen during WWII, so there was plenty of iron material to keep the mag busy, too much in fact. It was during this summer that I met Jon Christiansen and Les Savege at the dock in Fort Pierce. Les owned a diving company and had a good assortment of gear at his business in Orlando. Jon was the captain and owner of the *Capt. Jack*, an old 72-foot wooden Coast Guard buoy tender. It was a heavy, sturdy boat with a single 671 Detroit diesel engine for power. Jon was from the Newport News area of Virginia. He was born in Norway and could read Norwegian and speak

The Capt. Jack at the fuel dock

fluently in the Norwegian language. Captain Jon with his gentle soft-spoken manner had been on the water practically all his life and he earned my respect many times over. There are many individuals who may consider themselves captains because they have a license, but that doesn't make them a captain in my book. Jon is one of the few I have no problem calling "Captain".

Moving ballast stones at the Wedge Wreck exposed artifacts where they might be found. On occasion a jetting hose was used to clear sand as seen below where planking from the ship was still in plain view.   Photos: Jim Ryan

A diver is seen here measuring wooden rubble at the Wedge Wreck with a tape measure. Wood is seldom found on the 1715 Fleet wrecks today, having finally disintigrated in recent decades. Photo: Jim Ryan

Above: the panther figure recovered by Kirk Purvis   Photo: Tom Kraft

*Below: a toy animal, a rabbit, made of ceramic material was
only one of many toy pieces recovered at the Wedge Wreck.*

The days of summer passed quickly. It was late in August when San Francisco Kirk surfaced with a ceramic figure of a black panther. It was maybe 12 to 14 inches long, mostly intact except for a few paws. We were in an area that soon was producing small ceramic figurines, human figures, little jugs, little bowls, and what appeared to be small ceramic toys (confirmed as such later on). I found a small rabbit figurine in the same area. In this area of the ballast pile, we were also finding semi-rare gemstones like sodalite.

The "Riddle Wreck" was in very close proximity to dry land. The approximate area of the excavation is near the center of this photo.

Two local men from the Fort Pierce area contacted Frank with a proposal for him later that summer. I can't remember their names, but I want to say one went by the name of Edgar. They had found items that lead them to believe they had found an old wreck in the river near Sewall's Point back in the Indian River Lagoon, which they thought might be a pirate wreck. We ended up calling this site the "Riddle Wreck". Now I must admit, whenever I hear the word "pirate" associated with anything, my skepticism grows. As I recall this brief episode now, I can hardly believe we didn't get arrested. Frank made a deal with them... we would spend a day investigating for them and maybe blow a hole or two in the area. Edgar and his partner met us at the marina, and off we went down the Intracoastal Waterway. We made our way to just northwest of the St. Lucie inlet. They directed us to go near one of a few islands just off the shoreline. We dove and looked around. The visibility was just a few feet, as usual. Jim came up and said he felt something round like a cannon, which piqued our interest. We all agreed we should explore further, and we hastily dropped the blower into the water. Now we just looked like any other boat in the area with no tell-tale blower assembly in full view. We swam the anchors a short distance, just enough to allow us to sweep an area as we were blowing away the overburden. We blew for a short time and Jim descended again. He came up immediately and asked for a line to tie to an object below. He resurfaced, and we heaved ho on the line below. The object was heavy, whatever it was. After a moment, the object broke the surface. It looked like a log which had one end chopped down to a point, like a primitive post surrounding a fort perhaps. It appeared it had been protected in the mud because the pointed end looked like someone had fashioned it with an ax. Jim said he was sure there were many more down below. I went down to investigate. It was pitch black, but I could feel that there were more pointed logs in the mud. As I was below examining the hole, Jim

Jimmy Ryan shows off a bottle found at the Riddle Wreck site.

was standing on the edge above me, when the side of the hole gave way and came tumbling down on top of me. I felt something hit me. A large clump of clam shells came crashing down on me. I still have a scar on my hand to this day from that pile of dead shells. It didn't stop me from a closer inspection. I soon surfaced with a round-bottom bottle embellished with a "#3" on its bottom. Jim followed with a black bottle in perfect condition. Edgar definitely had something here, but what? There were logs cut and pointed at the top, there were bottles from what looked to be around the early

Above: the author, left, and John Kimbrell examine the top of an olive jar from the Wedge Wreck.

Below: the author poses with various encrusted objects and iron artifacts recovered from the Wedge Wreck.

1800s. We also brought up a few unidentified encrusted objects of iron and there was more material down on the bottom. We decided that it was time to go because excavating in the river without a permit on a potential historical site could have translated into a big headache for all of us. We pulled up the blower and the anchors and make our way north, back up the Intracoastal Waterway like it was just another day at the office. What had Edgar and his partner found? What was it we uncovered and dove on that day? I'm convinced that it somehow relates to the second Seminole war around 1838, possibly one of General Jesup's supply depots, or perhaps... old Fort Jupiter?

Doug Norton, a medical software developer from Longwood, Florida, was one of Frank's investors. I had given dive lessons to Doug, his wife Joan, and three of his sons. Doug was eager to dive on a Spanish Galleon. One calm day that summer,

Wedge Wreck timbers cleared of ballast stones   Photo: Jim Ryan

Doug and his family came on board to spend the day on the site. Doug wanted to go down first before anyone else in his family to make sure it was safe. Doug suited up with excitement, anticipating what he might encounter below. Jimmy was Doug's dive partner as they descended to the large pile of ballast below. Jimmy gave Doug the nickel tour with a swim around the perimeter of the wreck site. I remained topside, watching Doug's wife and boys stare down into the cloudy water below. Bubbles appeared as Doug and Jimmy came to the surface and started shedding their gear. Doug's wife looked at his face with anticipation, "What did you see?" she asked. I would imagine she was hoping he would reply, "Gold and silver everywhere!"

Doug looked at her and said, "Rocks, lots of rocks." I could see her face change just like a kid's face changes on Christmas when he or she realizes they got a pair of socks instead of that favorite toy. It was the same look. As Doug continued to remove his gear, I could hear Joan giving him a hard time under her breath about... "There goes our new washer and dryer." The boys took their turn diving on the ballast pile and came away with the same destroyed expressions. It was a big pile of rocks. I could see the wheels in Doug's mind turning that day. He wasn't one to give up easily.

As the days grew shorter and the winds stronger, our salvage season was almost over. We made no spectacular finds. We recovered lots of "E.O.'s" (encrusted objects), many pot shards, pieces of china, and small figurines. We knew next season we needed to get rid of the *Ocean Runner*; that we knew for sure. We had made it through the season, but it was time to move up and on. The lease on the apartment was over, but the stain on the kitchen wall from Kirk's hairdressing would remain.

Luckily, we had a new gig for the fall. Doug cut a deal with Frank for the rights to the ballast rock. Doug then turned around and hired us to bring up the ballast rock, billed by the ton... he then cut a deal with Neiman-Marcus, an American chain of luxury department stores owned by the Neiman Marcus Group, headquartered in Dallas, Texas, to sell fireplaces in their catalog made from authentic ballast rock from a Spanish galleon. Brilliant!

Pieces of Kang Hsi porcelain found on the Wedge Wreck are seen here, similar to those shards found on the other 1715 Fleet wrecks. Seldom are any intact pieces found today. These porcelain pieces originated in China, and were brought to Mexico's Pacific coast, then loaded on mule trains that carried them across Mexico to Vera Cruz, where they were loaded on another vessel for transit to Spain via Havana, Cuba.

Chapter 12
You Don't Look Like A Clammer

After the season shut down, we went to work stripping down the *Ocean Runner* for her new job hauling ballast stone. The Volvo engine we removed and resold for the company, no problem. Frank bequeathed the hull to us. We added a small lifting davit to the boat and some pulleys. I scrounged around Port Canaveral and found some discarded shrimp net, from which we excised a square section. This rig would be used as a means to lift the ballast stones. We would then take the four corners of the net and hook them together, and haul up the load. We cut a deal with one of our commercial fishing buddies at the marina to tow us out to the site. We would anchor up with several anchors and bring up the rocks. He would come back at a predetermined time to tow us back into the inlet. We also got our young buddy Glenn to stay topside and hoist the stones to the surface when we tugged on the buoy indicating it was ready to haul up.

Once we unloaded them onshore we would rent a U-Haul truck, get an empty weight on it, then load the truck up and reweigh it for our total. Doug then paid us for the weight on the ticket. Often we got suspicious looks when we returned the truck smelling like seaweed. We even had one dealer who became so suspicious he refused to rent to us anymore. As we got deeper into the fall season, the winds increased and the temperature dropped. Some days we would freeze our butts off, stuck out there waiting for our tow back to port.

We were working one blustery day and I could see a sailboat tacking back and forth, trying to make her way close to us. She wasn't making much progress, but after a couple of hours, she was close enough to yell over to us. A guy on the bow yelled, "Hey, how about a tow into port?"

I yelled back, "We haven't got a motor."

"What?"

I yelled back again, "I said we haven't got a motor."

He looked totally confused as he drifted closer. I explained the situation to him. Our best offer was for him to anchor up and when our tow came to get us, we could add him to the chain. That is exactly what we did. Our tow back to port pulled us with the sailboat in tow behind us. That was a weird sight for anyone at the inlet that day. I remember many days it being so rough going out the inlet we tied buoys to our gear bags in case we went down making the passage. We always had our wetsuits on, ready to go over the side if we had to jump.

It was freezing cold, and I always had a thermos of hot Campbell's Chicken noodle soup with me. I remember Alex coming up out of the water on one occasion and I poured him a cup. "Oh my God, what is that, it's so good?"

"It's just Campbell's canned soup," I said.

"Really, that's all, it's great!"

It just goes to show that an immediate perspective coupled with your environment is everything in a given moment. The weather eventually prevented any more diving, and the rock recovery ended. I confess, I'm not very proud of this endeavor, but it was all legal.

At this time, there was a boom in the Florida clamming in-dustry brought on by over-harvesting and devastation of the industry up north. The price for a bushel of clams skyrocketed and you could make a hundred or two hundred bucks a day wading around with an inner tube in the right place in the river. Clams were abundant to the point of being a nuisance for marine contractors who might be bury-ing cable or pipe on the bottom of the river. The area was flooded in no time with unemployed clammers from up north. All their trucks had plates from New York, Virginia, etc. Driving down to US 1 in the vicinity of Micco and Sebastian you would look out across the Indian River Lagoon and see hundreds of small boats raking up the clams as fast as they could. I remember standing in a line at the Marine Patrol headquarters in Titusville to get my clamming license. All that was required for proof of residency was an electric bill or utility bill in your name. The line was long, and the accents were all Yankee accents. I was wearing a Dap hat, which a friend had given me. When I finally got to to the registry clerk I provided my documents and answered her questions. She looked up suspiciously and said, "You don't look like a clammer." I took that as a compliment after surveying the surrounding crowd.

The author mugs for the camera here, seen in his "Dap" hat, with a knife firmly clenched in his teeth. He's holding a copy of "Pieces Of Eight", by Kip Wagner, a testament to every treasure hunter's dream of recovering sunken treasure. Photo: Tom Kraft

After the northern clammers had raked up all the clams they could with a rake, the law allowed them to dive for them in grassy areas. It didn't take long to wipe the clams out. This was terrible mismanagement of our resources and the over-harvesting was driven on by pure greed. Without the millions of clams to filter and clean the water, the lagoon's waters slowly became cloudy, no longer allowing light to get to the sea grasses. When the grass died off, the tiny shrimp and juvenile fish had nowhere to grow. It was the beginning of the end for the river. The politicians and environmentalists can throw all the money they wish at the Indian River Lagoon but it will never be the same, not even close. If everyone moved back to where they came from, the lagoon might have a chance. You can't prostitute out

The USNS Redstone
Courtesy of Wikipedia

your environment for growth and money and expect her to keep her virgin qualities. What you see when you look at the lagoon now is a gasping body of water. It breaks my heart.

I ran ads in the papers for diving jobs, hull cleaning, recovery of lost items, you name it. I got a call from a mutual friend at Port Canaveral who needed a diver for a job he was doing at the port. David Sands asked if I'd help him with an inspection of the water intakes on the *USNS Redstone*. She was 523 feet long with a 68-foot beam. When I arrived at the port, David greeted me and took me for a walk to the side of the ship where there was an intake symbol over a hole that looked to be about two and a half feet in diameter. "Randy, you don't mind tight places, do you," David asked.

"Sorta depends David," I answered with a grin. David explained that the hole we were looking at was one of several intakes for the engine room, and something was clogging the intake. The intake looked to be about 3 to 4 feet below the waterline. My job was to crawl in head-first with my hands out in front of me, which was the only way because it was such a small opening. This presented a couple of issues. The first issue was there was no room for the diver to wear a "bailout" bottle, which was a SCUBA tank on his back for emergency air supply in case the surface-supplied air failed. The second was the extremely restricted movement, or the lack thereof. I could crawl forward a little at a time, but not back up. Somebody would have to drag me out by my feet. David used "Kirby-Morgan Superlite hats", a good hat with acceptable diver communications built-in[1].

So the plan was that I would crawl with my hands in front of me, pulling myself along and grabbing any obstructions, then tell David to pull me out… sort of like me being a human grappling hook of sorts. I suited up, everything checked out, and I submerged. I could hear the rather loud hum from the generators running inside the vessel. The water was pea green, as usual, I extended my arms into the hole and dragged myself into the intake. I went a few feet and stopped, I asked David to pull me back, he pulled and I slid back towards the entrance. Good! I continued to crawl and after maybe a distance of 10 feet inside the pipe I saw the problem... plastic garbage bags. Most likely they had created their own problem by throwing plastic bags of garbage over the side of the ship at sea. Karma had caught up with the *Redstone*. I grabbed a handful and told David to drag me out. David pulled me and the garbage out. I crawled back in again. We repeated this exercise maybe a dozen times before we had the intake cleared. That little job kept me in groceries for a couple of weeks.

Frank Allen had been in touch, and we started discussing our next season's work on the Wedge Wreck. Captain Jon knew of a couple of boats for sale in the Hampton Roads area where he lived in Virginia. We started talking about living on one boat or using it as a base, while we used the smaller one for day-to-day work. The first boat was an old wooden 44-foot cabin cruiser with a single screw

1 Commercial divers use surface-supplied air wearing full-face masks, or helmets, which they commonly refer to as "hats". There are a number of "hats" on the market, with the Kirby-Morgan brand being the most popular. Helmets are equipped with earphones and microphones, as are the full-face masks, sometimes referred to as "BandMasks".

inboard. The second boat was a wooden boat, about 30 feet long with a straight inboard engine. It was a decent little work boat. Jon persuaded us to bring Frank up and look over the vessels to see if we could cut a deal so Frank, Jimmy, and I flew up to Norfolk to look at the boats. Frank had a terrible cold and was hacking and coughing the entire trip. He insisted on showing us the yuck he was coughing up in his hanky whether we wanted to see it or not. Frank was pushed around by a skycap while Jim and I walked close by, trying to blend in with the background in the Norfolk terminal.

Captain Jon had come to pick us up and as we got in the car, the skycap stuck his hand out for the tip which Frank promptly grabbed and pumped up and down thanking him for his wonderful service. Jimmy and I about died. We thought that was the funniest thing we'd seen in a long time! "Classic Frank" we found out with more to come. We made our way to Hampton Roads. Hampton Roads is a coastal community along a body of water known as Hampton Roads, which is one of the world's largest natural harbors (more accurately a roadstead or "roads"). The area is home to hundreds of historical sites and attractions. The harbor was the key to Hampton Road's growth, both on land and in water-related industry.

Jon showed us the two vessels, and we all agreed we would be much better off with these two vessels than we were the previous year with the *Ocean Runner* and an expensive apartment. The smaller boat was the *Bryan Michael*, the larger boat we renamed the *Frank*. We headed back to the airport for the trip back to Florida. On the way we watched Frank drop a gold coin in a hand or two, always asking his audience if the height above their hand was six inches. He'd then drop the gold doubloon in their unsuspecting hand. He was a master at getting people's attention with that gold coin. During the flight back, we decided we would return to Hampton Roads in early spring to bring the boats back to Florida.

Chapter 13
Greenhead Flies, Smugglers & Hot Mustard

Much to my disappointment, the winter passed with no beach cuts. I started calling the beach "the Bank of Spain", hoping that it would open with every passing low-pressure front. I could use a withdrawal. But the bank rarely opens, sometimes it will stay closed for years. Spring was right around the corner, and fortunately, it was not long before Jim and I were on an airplane headed for Norfolk to join Captain Jon on our vessel transport adventure. We would be bringing the boats back to Florida via the Intracoastal waterway, or the "ditch" as many refer to it.

The Intracoastal Waterway (ICW) is a 3,000-mile-long inland waterway along the Atlantic and Gulf coasts of the United States, running from Boston, Massachusetts, southward along the Atlantic Seaboard, around the southern tip of Florida, then following the Gulf Coast to Brownsville, Texas. Some passages in Virginia date back to the 1700s and are still being used today by modern mariners.

We towed the smaller vessel, the *Bryan Michael* with the larger boat, the *Frank* much of the time. Sometimes conditions forced us to put a man on the smaller boat in tow. We departed Hampton Roads full of excitement as we passed the hundreds of moth-balled military vessels docked at Norfolk naval base. As we passed, I recalled the paintbrush incident with Skipper years earlier when I was diving for Aqua-Tech and a smile came to my face. We soon entered one of the old channels that ran right next to a small highway. It was strange to have cars passing by only 30 feet away from us as we made our way south down the old waterway. We continued down the old skinny canal until we emerged into a wider section of the Intracoastal. At nightfall, we found a small marina and docked for the night.

Our trip would take us close to a thousand miles down the east coast, passing through Morehead City, Myrtle Beach, Charleston, Beaufort, Savannah, and then passing Fernandina Beach into Florida. This route is made up of interconnected rivers, bays, channels, and sounds as you make your way up and down the coast in much calmer waters than those offshore. We casually made our way from one small seaside town to the next. The scenery of the old south along the way was full of history. We cruised by shanties, plantations, large mansions surrounded by immense oak trees, and high-end marinas full of million-dollar sportfishing boats. Passing us we saw crabbers, shrimpers, sailboats, fuel barges, sport anglers, and pleasure boaters of all sorts.

As we made our way into the Carolinas we were attacked relentlessly by a most hideous adversary that continually distracted the skipper's navigation. The enemy... Greenhead flies, nasty biting, blood-letting, swarming, vicious vampire bugs. They flew into the cabin one after another, feasting on us while we cursed the little bastards with all the terrible names we could conjure up. Jim and I armed ourselves with rolled-up newspapers fashioned into fly swatters and stood guard over Captain Jon, so he could continue to operate the boat without dangerous distractions. Soon, their little smashed bodies littered the deck. Seeing them dead, our defense became all the more invigorated. Anyone who has been through this area in the spring will chime right in when you talk about Greenhead flies. A couple of store-bought fly swatters were on our shopping list at our next layover for sure.

That evening we pulled into the only marina at hand which was normally way out of our price range, but we had no choice. All around us were Hatteras, Viking, Post, and Bertrams. We were way out of our league. The marina's cocktail hour surrounded us with laughter, and conversation filled the air. I walked to the shower and on my return along the dock, a southern bell on one yacht called me over and asked me if I would take a rather large platter of chilled blue crab claws off her hands as she had way too many. I graciously accepted her offer, thanked her profusely, and skipped on down the dock with the crab claw platter, feeling like I'd just won the lottery. The guys saw me coming and began asking questions about where I had scored such a feast. In their eyes, I must have the Midas touch. "Just pure good luck", was my answer. I suggested they now do their part and drop in on the marina store for some malted beverages to consume with our blue crab claws and horseradish dip. Life was good. We were in high cotton, so to speak. We devoured the blue crab claws until we couldn't eat anymore; it was an enormous platter. We washed them down with cold beer and it wasn't long before we dreaming of a land devoid of Greenhead flies.

We awoke the next morning refreshed and once refueled we headed south again, getting ready to cross into South Carolina, hopefully, later that day. We were just above Pawley's Island tied to a dock as we took on more fuel when Captain Jon had the engine hatches propped open, checking oil levels. I heard what sounded like a speed boat running at high speed; I looked up and sure enough, here came a cigarette boat all-out flying down the river, with a Coast Guard patrol boat right on its ass. Captain Jon heard it too, "What's that", he asked. Before I could reply both the smuggler and Coast Guard boat flew by creating a 3-foot wake as they passed. The wake hit us and the engine hatches came crashing down on top of Jon's head, knocking him down into the engine compartment. The smugglers headed for the riverbank, running aground, jumping out of the boat, and fleeing through the woods with the Coast Guard crew following them through the woods. Jon came up out of the engine compartment holding his head and asking, "Who did that?" I pointed at the Coast Guard boat now tied to the end of the dock with two crew members remaining aboard, while the others were in hot land pursuit. To say it pissed Captain Jon off would be an understatement, a colossal understatement. Jon let go on those Coast Guard guys with a merciless verbal assault that would bring a blush to the most hardened old salts. I thought we'd be on our way minus one for sure when Jon let go on them. They took it... they knew they had placed Jon in danger along with many others in their high-speed pursuit of the smugglers. They graciously took a superb ass-chewing, filled out a report on the accident, noting damage to our gunnels from getting slammed against the dock. The incident slowed us down a little, but we hoped to make it to Charleston by the next evening with no further incidents. That was not to be.

After the excitement of being rocked against the dock by smugglers and the Coast Guard, we were soon southbound headed towards Charleston where we intended to dock for the night. We came by McClellanville, a small sleepy fishing village surrounded by century-old oak trees, and best known for its shrimping fleet. We had no way of knowing at the time but in 1989 the town would be devastated by the full brunt of Hurricane Hugo which would destroy homes, downing the many old oaks, throwing shrimp boats in front yards, and otherwise altering much of the picturesque character of this historic fishing village. The stronger north side of the eyewall passed directly over the village as a Category 4 hurricane. Residents taking refuge in the local high school, a designated storm shelter, were surprised by a storm surge that threatened to drown them. Helping one another in complete darkness, they crawled into a space above the false ceilings of the building and all of them survived. It would never again look like it did that day when we passed by on our journey south.

We slowly cruised by the Isle of Palms, where Lee Spence, as a curious youth, had located many shipwrecks with his research and local knowledge. Just offshore lay the remains of many shipwrecks including the *Stonewall Jackson* of 870 tons, which ran ashore in eight feet of water and burned, never salvaged. This area was first settled in 1670 and is rich with the history of conflict. We passed Sullivan's Island, reminding me I was to get in touch with Lee Spence as soon as we docked in Charleston to meet him for dinner. As we passed Sullivan's Island we slowly proceeded across Charleston harbor headed towards the popular downtown marina and dock. I glanced back to see a US Customs boat right alongside us. They came out of nowhere and were climbing aboard as we drifted to a stop. This turn of events rather confused and somewhat amused us, as we had never left the Intracoastal Waterway since we departed. The customs agents started asking questions about where we were coming from and what we were doing. It was hard for us to take them seriously as we answered their questions, trying not to smile too much. We soon convinced them we were not smugglers, but we had seen a smuggler and suffered the consequences of the chase as we pointed to our crushed gunnel. Our lack of seriousness did not amuse them, and they eventually climbed back aboard their vessel with less enthusiasm than when they first climbed aboard ours. We continued to scratch our heads, wondering where they might have gotten their motivation to board us. The sky in the west was glowing pink when we pulled into the Charleston City Marina to tie up for the evening.

Charleston is a fun town, especially if you enjoy history and southern cuisine. Charleston was frequently the target for pirate attacks in the early days of the settlement. In 1718, none other than Edward Teach, aka Blackbeard, attacked several ships trying to enter the Charleston Harbor. He took hostages and ransomed them for a chest of medicine. Walking through the streets of Charleston you would never imagine that on August 31, 1886, the largest earthquake ever recorded in the southeast occurred near Charleston. A remarkable 7.8 magnitude quake rattled the area, and it damaged buildings

The author at the wheel of the Frank, navigating the ICW

The author, Randy Lathrop at left, and Lee Spence

in states as far away as Ohio. It killed 60 people, caused more than five million in damages, and about 14,000 chimneys were destroyed. Lee Spence met us for dinner. He pretended the bowl of water and slice of lemon placed on the table to cleanse your fingers was lemon soup, complaining it was watery was hilarious. It was hard to keep a straight face seeing the expression of disbelief on the server's face. He must have been astonished that anyone could be so unrefined.

Lee had spent most of his youth here and was well known in the community. You can buy his books and maps at the tourist info center downtown, and it was fun having him as a guide. We spent the evening telling stories and hopping from one drinking establishment to another.

Lee took us to a bar that served Lee's "hot mustard". He wanted us to try it, as it was his original recipe. A word of warning; if Lee ever asks you to taste something, beware. He's good at keeping a straight face until it passes over your lips, then he'll break into a big grin as you begin to suffer. Hot, hot, hot! And it lasts for a long time! Lee eats it like ice cream, so don't be fooled. We talked about working for Frank and our plans for the summer. Lee asked if I could put in a kind word for him with Frank if we needed another hand. Considering Lee's credentials I was sure he could be an asset. I promised Lee I would do that as I enjoyed his company and would like to have him on board for the summer. We slowly made our way back to the marina and said our farewells. Incidentally, the tides in Charleston are 5 to 6 feet, so tie-up with that in mind. When we returned our lines were guitar string tight, but we made it just in time to adjust them for the tide levels and we slept soundly that night.

We had new scenery every day as we made our way further south down the ICW, past Hilton Head, Tybee Island, and St. Catherine's Island into Georgia. This part of the trip was interesting to us, as it was much less developed than what we had previously seen on our trip. We marveled at all the ballast stones we saw piled here and there on small islands and along the shoreline. All rock and stone ballast, an assortment for sure, but much of it looked to be square-cut granite. This was another section of our southern coastline saturated with history from pre-colonial times to the Civil War, and we could see it and feel it around us. We passed through the town of Thunderbolt, Georgia, population of 2668. The little town is known for its picturesque atmosphere and seafood restaurants which draw many visitors. It's also known for something else I'm told by reliable sources. I know a former smuggler who served time in the federal prison system for dealing in the Devil's Lettuce, and he told me this story. The Federal Bureau of Prisons will contract with smaller local jails to house their inmates while they move them around the country from court to prison or prison to prison. Most of these places throw a bologna sandwich at you and call it a day, their job done. But a rare few are much different, the jail in Thunderbolt is one of these rare ones. Practically all the federal inmates they watch are low security, nonviolent drug offenders. Thunderbolt, it seems, is an authentic Mayberry type jail, complete with southern home cooking. The jail uses a couple of church women to cook for the inmates and those gracious ladies just cook the only way they know how, for family.

Captain Jon Christiansen, captain of the Capt. Jack and the Heather Glynn.

Inmates throughout the federal prison system know the name of that little jail, and it was the best thing that could happen to an inmate's stomach for a long time. If you landed in that little jail in Thunderbolt you won the lottery and were in for some fabulous home cooking.

The scenery just kept improving as we kept our bow pointed south for Florida. From Savannah to Little Saint Simon's Island there were more ballast stones along the shorelines... unending piles of ballast. Our next stop would be Fernandina Beach, Florida, on Amelia Island, known as the "Isle of 8 Flags". We docked at Fernandina and made a beeline for Florida's oldest bar, the Palace Saloon. Louis G. Hirth purchased the building in 1903 and turned it into a bar. Sitting at the hand-carved mahogany mirrored bar sipping on an Old Fashion, I traveled back in time. I could imagine patrons, including the Rockefellers and Carnegies, sitting underneath the molded red and gold tin ceiling. The hand-painted murals on the walls gave the overall regal ambiance that must have impressed the regulars back in the early 1900s. After some lively conversation at Florida's oldest watering hole, we staggered back and slept soundly. We threw the lines off early the next day and headed to St. Augustine, which was the last city marina we would encounter before our final destination at Fort Pierce.

We sailed the same waters as had the French back in the 1500s while passing through the Jacksonville area. Lucky for us we didn't have a Spanish fleet on our tails in hot pursuit like the doomed Frenchmen had centuries ago. We docked in the "Ancient City" and soon set out in a town I knew very well. I remember staying at a boarding house in St. Augustine in 1974 for eight dollars a night. The lady who owned the large 3-story home had no recourse but to rent rooms to make ends meet. The house was old and run down and every day as the owner began drinking she would retire and lock herself in her bedroom. She also kept a lock on the kitchen but one night that didn't stop a hungry thief from removing the pins in the door hinges and gaining entry to the refrigerator.

The town had changed since 1974. It was more touristy now, much of the charm replaced by rampant "Oldest City" commercialization. I used to drink at the Monks Vineyard, now closed. My favorite watering hole in the old city has to be the Trade Winds Bar. Once known as the South Seas Lounge, it now appeared to be a sailor bar. The Trade Winds is full of many island artifacts and collectible items placed all around. Many of them still hang to this day. It was here, back in 1974, that I met Richard Boone sitting at the bar. Richard Boone was famous for his hit TV series, "Paladin", also known as "Have gun will travel". The dark bar provided refuge on a hot sunny day and I took a seat next to another patron. We struck up a conversation about nothing in particular. We drank and shared stories over several rounds, he always insisting to buy the drinks. A couple of hours went by and my new friend wishing me a safe voyage and much luck in my life as he departed. After he left, the bartender asked me if I knew who that person was I had just spent the last couple of hours talking with. I said, "No, I'd just now met him."

He said, "You remember Paladin?"

"You mean the guy from Have Gun Will Travel?"

The bartender answered, "Yes, that's who you've been drinking with for the past several hours."

My new best friend I'd been drinking with was Richard Boone, a movie star in such films as "The Halls of Montezuma", "Return of the Texan", and "Man on a Tight Rope". I hold that day very special in my bank of memories. I'm only bummed I didn't get a "Have Gun Will Travel" business card as featured in the TV show. I would eventually return to St. Augustine 35 years later to marry my wife in a sunset ceremony at the Castillo de San Marcos. So St. Augustine will always hold a special place in my heart.

We greeted the morning knowing it was our last day of this voyage down the ICW. We were sad it was ending, but excited about the times soon to come. Just south of Daytona we passed New Smyrna where the ruins of a failed enterprise by Dr. Andrew Turnbull in 1768 still exist. A little further on we passed Ross Point, the terminus of Ross Road, which can be seen on many of Florida's oldest maps. The Canaveral coastline, littered with ancient shipwrecks and modern rocket debris all mixed in together on the seafloor was just a few miles away as we finally cruised into my stomping grounds. Oak Hill and points south were my backyard. I have shrimped Oak Hill for years, camped at most of the islands, exploring by foot, canoe, and car. Whenever boredom would raise its ugly head, a trip to Mosquito Lagoon would be a sure remedy; it was always an adventure. Most logical folks would want nothing to do with the swarms of mosquitoes, the strong summer squalls, and the "music worms" (rattlesnakes) lurking in the underbrush. And let's not forget to mention the alligators, the ones that aren't supposed to be where you find them.

I was canoeing Turnbull Creek on a cold winter's day. The tide had dropped and the little creek had narrowed to maybe ten feet across because of the low water. I slowly came around a corner next to a high bank when I came face to face with a solid 12-foot alligator on the bank. We were at eye level with one another. He looked at me and I looked at him. It seemed to register in both our minds at about the same time as to what each of us was seeing. He responded first by launching himself off his perch only a few feet in front of me. My canoe rocked up in the air as he leaped into the shallow water, almost hitting the front of my 14-foot canoe. This gator didn't understand that it was too cold for gators, and he wasn't where he should have

The west bank of Mosquito Lagoon, part of the ICW, is the home of Ross Hammock, seen here. There's several Indian Mounds nearby.

been that day. I lost an almost-fresh pair of skivvies that day, and we'll just leave it at that. Just understand gators don't care what the experts say, they'll go where they want to go. Saltwater or freshwater, trust me, they don't mind.

Just a few miles to our east was the Canaveral coastline with 400 years of maritime commerce gone bad lying on the ocean bottom. You can find some celebrated wrecks there but many remain undiscovered. Some are found and never spoken of. Mysteries abound along the Canaveral coast.

We slowly crawled through the narrow Haulover Canal, then south past the ruins and ghosts of Fort Ann. The Haulover was a strategic portage place used by Indians and traders to move from the Indian River Lagoon to the Mosquito Lagoon where they might finally access the open sea via Mosquito Inlet at New Smyrna. Fort Ann was to serve as a garrison post for United States troops as a depot for supplies and equipment in support of General Thomas Sidney

Jesup's military offensive against the Seminoles in 1837-1838. The fort was established in December of 1837 and used through April of 1838. The fort played a significant role in the operations against the Seminoles, supplying troops in the field and other forts. In April 1838 it was ordered abandoned. Today it takes a trained eye to distinguish any remains of its earthen works in the thick brush indicating where it existed almost 200 years ago. Fort Ann is left to neglect over the years with little excavation or any studies performed on the site. The government will spend huge amounts on lawyers and legal fees, enforcing their sovereignty and regulatory oversight, squandering money better spent to do something useful. No, you, the average citizen, may not trespass upon the ruins of Fort Ann while with each passing day nature takes back a little more, piece by piece, till the fort is lost forever.

We motored by Titusville and Kennedy Space Center where the launch pads and Vehicle Assembly Building (VAB) were clearly visible to our east on the horizon. From Titusville southward to Fort Pierce the river widens, but the channel is the only part with any depth. The crab pots are many on the river here, unfortunately probably the last fresh seafood you can get in this area these days, and that supply is fading fast. Of course, the term "river" is a misnomer. It's a lagoon, with no flow, except wind-driven. It's a saltwater lagoon for the most part, only brackish in very specific areas where freshwater mixes with it. The salt content of the lagoon fluctuates up and down, and sometimes it's saltier than the ocean itself. Known by several names over time, Rio de Ais by the Spanish, then referred to as the Hampton River by the English during their occupation, it eventually became known as the Indian River, owing principally to its original name of the Rio de Ais.

We were now on the home stretch past Melbourne, then south past Sebastian, where we recalled our time with Chuck Kenworthy and Quest Exploration as we passed the location where the *Quo Vadis* and the *Searchers* once docked. As we motored by Vero Beach, I thought of my old friends Bill and Jan, Diveco Diving Systems, and how much I owed both of them for setting me on this path. It was cocktail hour when we pulled into Mel Borne's Marina, that's Mel as in Melvin, and secured the lines.

Chapter 14
Why I Hate Boatyards

The salvage season of 1981 started with new boats and new crew members. John Kembrell had stayed with Perry Submarine this season, and Tom Kraft went back to his job at the photo lab. We gave up on any more magging because there was too much trash in the area, so Kirk Purvis stayed in San Francisco. The main crew remained: Jim, Alex, and myself. We had a few new divers come and go, diver friends who helped on weekends. I spoke to Frank about Lee Spence, and Frank placed Lee on the payroll. The larger boat, the *Frank* stayed moored to the dock and served as our live-aboard headquarters. So living onboard the *Frank* were Jimmy, Lee, and me, plus my cat "Fluffy". The cat was a rescue animal and came to us with that name, so don't give me a hard time about the name. Captain Jon had his boat, the *Capt. Jack* tied up in Fort Pierce. He was looking for work, having just come off a job so he could join us occasionally. At the end of the last year's salvage season, we removed the blower from the *Ocean Runner* and needed to re-fabricate it for the new boat, the *Bryan Micheal*. We had dropped it just offshore of the site the previous year because we had no other place to store it. It held up very well laying on the bottom for seven months. We marveled over the excellent condition it was in. It had traveled maybe 50 feet over the course of the winter, cutting a deep, perfect path on the barren bottom. During the refitting of the *Bryan Michael*, we picked up another crew member, Tim. Tim was your typical jack-of-all-trades and master at none. Hanging out at the boatyard had served him well in

The Bryan Michael

picking up new gigs. He was a big guy with a friendly manner, and a Bob Dylan-loving, reborn Christian.

Once we had the blower refitted on the *Bryan Michael*, Tim talked Frank into a major overhaul on our big live-aboard boat the *Frank*. This took us from the cool ocean breezes of Mel Borne Marina to the stifling dead air of a polluted boatyard. We spent almost the entire month of June and part of July in the boatyard. Climbing up and down a ladder in a dusty boatyard many times a day can get old fast. It was hot and miserable working in the boatyard, living in the boatyard, even drinking in the boatyard sucked. Tim went home to his air-conditioned house every night while we sweated it out at the boatyard.

I learned a lesson that summer that has stuck with me since. I was walking to the restrooms one grueling, scorching afternoon and passed an old salt crouched under his sailboat seeking some shade from a bottom paint job he was doing. I looked over at him and quipped, "This is a bunch of shit, huh?"

He looked up at me, and answered, "I don't see any anchor tied to your ass, Sonny."

I thought about that and he was spot on. Sometimes people complain and find like complainers to chime in. This old salt wasn't having any of it. He did not find complaining benefited any situation.

After a few grueling months in the "Cracker Boy" boatyard, we finally made it back to the marina. We were back on our poor diver's diet, grouper head meat with pasta in tomato sauce washed down with Black Label beer, which we could get for a buck and ninety-six cents at the 7-11. After all that time in the boatyard, we didn't do much digging on the site that summer. We mostly moved the ballast around from one area to another to see what might lie underneath. We recovered terra cotta pottery shards, animal figurines, Kang Hsi porcelain, musket balls, cannonballs, ship spikes, lead sheathing. Real Eight corporation reported finding some similar animal figurines along with 13 silver wedges in 1960 when they worked the same wreck. We had yet to find any precious metals or coins. We moved a lot of ballast rock, exposing much of the keel, which was amazingly solid underneath.

Several incidents that summer come to mind. The first one was sort of funny and dangerous at the same time. Frank had come by two Farallon MK VI Diver Propulsion Vehicles. These were some of the first DPV's available to the public. The units were 55 inches long and weighed 85 pounds. You would lie on them with your arms extended. They were bright orange and looked like a torpedo. We couldn't wait to try them out, so we tried them out in the inlet directly off the boat dock. I still had my baggies on, so I stripped off my t-shirt, grabbed my mask and snorkel, and dragged one into the water. The thing was heavy and not neutrally buoyant. I laid down on it and took off. Once you got going, it was great. I was zipping up and down like a dolphin. Diving deep then coming up for a breath of air. These things were so much fun. Jimmy had grabbed the other DPV and was doing the same. I got a good breath of air and checked out the bottom of the inlet maybe 40 feet below. I zipped down to the bottom and started to come up when I felt the drawstring on swim trunks being sucked into the little propeller on the DPV. The unit stopped, and I

A view toward the stern of the Bryan Michael in the vicinity of the Wedge Wreck

was stuck on the bottom attached to an 85-pound DPV anchor. It had me pulled in tight; I struggled to get my swimsuit string unwrapped from the propeller. It wasn't coming loose, and I was still holding my breath, my "only" breath. If I couldn't get the string unwrapped, my next choice was to leave my swim trunks and the DPV on the bottom. That would be embarrassing, coming to the surface bare-ass naked without my shorts and the DPV. I wouldn't live that one down for a long time. I continued to struggle with the string, and finally, I got it unwrapped. Hitting the trigger on the DPV, I headed up, looking like a Polaris missile when I broke the surface gasping for air. I immediately shared my experience with the others about loose, dangling swimsuit strings.

Frank Allen was a fair man. He told the crew the first year that they could keep the first silver or gold coin they found. After moving and removing much of the ballast rock now for two seasons, new items were coming to light. Late in the season, a single silver wedge was recovered by a member of the team. Real Eight Corporation had documented, finding 13 of these silver wedges from their time salvaging the wreck. The individual pieces when placed together made a round circle. These silver wedges were most likely smuggled contraband silver concealed in the bottom of a barrel to avoid detection. That is the reason given why they formed a circle when placed together We were all thrilled for the find of the silver wedge; we were happy for Frank, he finally had something to show his investors. But the joy soon faded. Frank had told us from the beginning that the first "piece of eight" a diver found it was theirs to keep, whether it was gold or silver. The diver who found the wedge morphed this statement into the "first piece of silver". Frank Allen had always been kind and generous to us, his success was our success. The rest of us felt terrible about this, we knew better. It didn't get any better with that diver being difficult about Frank taking possession of the wedge to show his investors. Frank could have possession for a short time, but the silver wedge ended up with the diver. Frank could have, and should have sent him packing. But Frank was a stand-up guy. In retrospect, the rest of us should have grown a pair and stood up for Frank. I regret that even now.

We spent the last days of the summer salvage season conserving the artifacts in baths and storing them. Captain Jon had been working on a job for his boat the *Capt. Jack*. He had the storage holds insulated and fitted for hauling produce from the Dominican Republic to Miami and he had planned on leaving after the peak of hurricane season in September. Jimmy had hired on as his first mate, so he had a gig after Frank shut down for the season. The rest of the crew went back to their full-time jobs. I had no plans, but I was born to have an adventure.

The author shows off a catch of Spiney Lobsters aboard the Bryan Michael. Fish and lobster were an expected addition to daily fare while working the waters of the Treasure Coast. Photo: Jim Ryan

Chapter 15

You Have A Collect Call From The Turks & Caicos Islands

The salvage season for East Coast Research was over until next summer. We didn't know it then, but that season was the last season for Frank and East Coast Research. The *Capt. Jack* finally threw her lines and headed out Fort Pierce inlet for Santo Domingo to take on the first load of produce. I went back home to Cocoa and tried to figure out my next move.

I had a lot of catching up to do with my hometown buddies. One buddy we called OP, (because he always bummed other people's cigarettes) worked on a shrimp boat and was eager to talk to me. When I spoke to OP, he told me that while dragging for shrimp, northeast of Port Canaveral, they had snagged some wreckage. The captain of the shrimp boat was a local guy who was also a diver. He dove on the wreckage and had recovered a hatch cover and a few other artifacts which were part of a nondescript display at the 76 Dive Center in Cape Canaveral. They were used as props in the dive store, courtesy of the shrimp boat captain. I frequented that shop, so I looked at the stuff on display soon after I spoke to OP. The hatch had writing on it that said, *Vera Cruz*, or something similar. There were also some glasses and plates in the display.

It was not long before Lee called me checking in, and I mentioned to him in confidence what I knew about this mystery wreck. The next day Lee called and said he had a potential identity for the wreck as *The City of Vera Cruz*. Lee, through the years, had compiled information on shipwrecks in select areas. He provided a few dates and off I went to the University of Central Florida library to examine microfilm files of the Charleston Newspaper and the New York Herald. These papers and others provided detailed eyewitness accounts, along with other pages of information. The passenger list, the cargo, the wreck, the survivors, the aftermath all came alive via those spools of microfilm. I made several trips to the library and came back with partial copies of the records which required me to scotch tape the pieces together to make them whole again.

The story I began to uncover was fascinating; *The City of Vera Cruz* embarked from New York on Wednesday, Aug. 25, 1880, with 29 passengers, 48 crew members, and a cargo of goods that included mailbags bound for Havana and Vera Cruz, Mexico. There were also crates of cutlery, firearms, toys, and corsets. There was even a fully assembled railroad car secured onto the deck. There were horses on board, along with a civil war hero who was buried, dug up, and reburied. One interesting little tidbit was the story of a New York jeweler moving his business to Jamaica. The manifest also showed a safe or two listed on the manifest. This was getting interesting.

I was ready to walk out the door to run some errands when the phone rang.

"Collect call from a Captain Jon, will you accept?"

"Yes, I'll accept," I said.

"Randy, it's Jon. How fast can you get on a plane and be in Providenciales?"

"Where is Providenciales?" I learned that it was an island in the northwest Caicos Islands, part of The Turks and Caicos Islands, a British Territory. Captain Jon told me he needed a dive master ASAP, and he would explain things when I arrived in the islands. This was a sudden and exciting turn of events. How did this come to be, I wondered? Captain Jon and Jimmy were supposed to be hauling produce, not chasing shipwrecks. I soon found out.

After the *Capt. Jack* had arrived in Santo Domingo, they were shaken down by customs. Ended up costing them a few cassette tapes, a ball cap, and a few other items which the customs officers seemed to fancy. If the customs guys handled it during the boarding inspection, Jon or Jimmy would offer it to them. "You like it, here, take it." The officials also required them to stay on the boat in port until Jon's contact came through with some required paperwork. So, they sat there in the port, sneaking off for a drink occasionally. The days passed with no progress, no contact showed up, no paperwork arrived. It wasn't looking good for the *Capt. Jack*, Captain Jon, or Jimmy. Captain Jon looked out across the harbor, not knowing exactly where to go next or what to do. This project was dead in the water, literally. Jon's eyes stopped scanning and locked on one large orange vessel tied to the dock across the harbor. He pulled his field glasses from their case and focused them on that large orange vessel.

Captain Jon couldn't believe his eyes! He knew that boat; he was once its master. The name on the side read *Heather Glynn*. She was a 110-foot Norwegian North Sea salvage tug, which Captain Jon had skippered in Norway. Now here comes the good part. Every written or printed thing on that boat was in Norwegian. Captain Jon and Jimmy walked over to the tug to check it out. Standing outside the wheelhouse looking out over the port was a man wearing a beret and aviator sunglasses, in green military fatigues. His name was Roger Miklos. Roger had a company known as the Nomad Treasure Seekers. He had just leased out the *Heather Glynn* for a project in The Turks and Caicos Islands. The vessel had come with a captain and crew, only problem was the crew expected Roger to pay their back pay owed them from their previous employer. Roger wouldn't pay, so the captain walked off, but the two-man crew stayed. As long as the groceries held out the remaining two crewmen stayed... in reality, they had nowhere to go. As fate would have it, Roger was having a tough time finding a captain for the ship because all her documentation and operation manuals were written in Norwegian. When Jon Christiansen showed up, Miklos had his problems solved!

In a matter of a few hours Captain Jon was the new master of the *Heather Glynn*, and Jimmy was the new captain on the *Capt. Jack*. Roger also needed a dive master and a new engineer. Captain Jon immediately called the former engineer on the ship and he then called me for the dive master position. In a matter of a few hours, all of our destinies had changed, and whether for the better or for the worse remained to be seen.

The Heather Glynn   Photo: Jim Ryan

As recently as 1964, Providenciales did not have a single-wheeled vehicle. Roads, water, telephones, and electricity were also nonexistent. When my plane touched down in 1981, there were a few rum bars and not much else. After an enormous investment on Grace Bay Beach by Club Med several years later, the island's first large hotel and casino complex opened in 1984 and touched off a development boom. The islands are with little flora being a dry, barren landscape, saturated with salt, having little visual appeal, aside from the turquoise waters surrounding them.

My jet landed at the airport terminal at Grand Turk, where I was to catch a small plane to Providenciales. My contact at Grand Turk was the pilot of the small single-engine plane I was to take to Provo on Providenciales where I would meet the company seaplane to take me out to the ship. We lifted off, rising above the surrounding clear blue waters of the Caribbean. You could see the ocean bottom in 50-feet of water. The diving here is world-class because of the coral reefs surrounding the island chain. In short order, we were circling the airport at Providenciales. The first thing I noticed were the many airplanes sunk in the surrounding waters of the airport. Single-engine, as well as twin-engine, a variety of sunken aircraft, met my eye. As we began our approach to the landing strip, I heard a loud beep, followed by another. I could see a little red light blinking on the console. I tapped the pilot on the shoulder who had his headphones on. "What's that?" I asked.

"It's the landing gear indicator, it's not down," he replied.

"Has this happened before?" I asked.

"No."

"Oh shit," I thought to myself. The pilot pulled a book from next to his seat and begin flipping pages. "Oh double shit," this isn't good. He soon found the page he was looking for.

"See that lever next to you?" he asked.

"This one here, yes," I assured him.

"Keep pumping the handle up and down, that's the manual hydraulic pump for the landing gear," he explained.

I pumped and pumped like my life depended on it. He then flew by the airport tower, where hopefully the tower officials could confirm if the landing gear was up or down. We flew by the tower. The pilot was on the radio with them. "What did they say?" I asked.

"They can't tell," he said.

"Oh, triple shit," I thought. The pilot then explained we would just have to land, keeping our nose up until the last minute in case the gear didn't drop down. We lined up for our last approach. Coming in low, I looked down at the airplane carcasses below me and said a quiet prayer. The back wheels let out a little squeal as we touched down, the pilot keeping the nose up until it dropped down on its fully deployed front landing gear. My heart rate slowly returned to normal as coasted up to the terminal building.

The terminal was a small building with open windows in one corner of the airstrip. Nomad Treasure Seekers also had a seaplane and pilot on its payroll, whom I was to meet at the airport. I didn't have to wait long as I heard a voice behind me. "Randy?"

"Yes," I replied as I turned to see a bearded guy maybe in his late 20's.

"Let's go," he said. We did a short walk to the seaplane close by that had escaped my attention a few moments ago. We loaded up my gear and took off, once again flying over the skeletons of sunken planes below as we gained altitude. We flew for a few minutes when the pilot said, "There she is."

I looked down to see two vessels on the calm blue sea below us. The *Heather Glynn*, the 110-foot North Sea salvage tug was a nice large ship with a highly elevated wheelhouse, and tied to her stern was the *Capt. Jack*, 65 feet, 38 gross tons, powered by a single 671 diesel. I could see people waving at us as we circled above, ready to land. We were soon coasting up to the *Heather Glynn* and climbing aboard. Captain Jon and Jimmy greeted me and told me to place my gear in the dinghy tied up to the stern. I would stay on the *Capt. Jack* so we would transfer my gear later.

Captain Jon and Jimmy told me the story of Nomad Treasure Seekers and Roger Miklos. In 1977 Roger Miklos, along with two treasure hunters out of Key West, had discovered an early 16th-century wreck on Molasses reef near Providenciales. The two other fellows, Olin Frick and John Gasque took credit for discovering the wreck in 1977. They had a dispute with Roger about the wreck, who supposedly had an old chart showing its location. Roger quietly organized his own expedition, which including signing agreements with The Turks and Caicos government, all unbeknownst to Frick, Gasque, and the British government. Roger, a former California cop, was quite the showman. He had rings on every finger; he wore two watches, one on each arm. He wore many gold chains and he had one large chain which had three ships hanging from it. He had many military-style uniforms, and he even had an admiral's uniform with scrambled eggs... all the bling. PT Barnum would have been proud of Roger Miklos. When Roger Miklos had flown into The Turks And Caicos in his leased private plane to negotiate his deal with the locals, they were in awe of the uniformed Roger and all his bling.

A third player from the Institute of Nautical Archaeology, hereafter the INA, a nonprofit group affiliated with Texas A&M

University, and an influential archaeological organization, heard about the wreck and they were claiming the site for themselves. Not understanding that the two sites, the Miklos site, and the Texas A&M site were actually the same site, the island government told Nomad Treasure Seekers to take possession of their site and to protect it. In the meantime, Mr. Frick and Mr. Gasque, whose firm, Caribbean Ventures, located in Key West, Florida, had filed a $100 million lawsuit against the INA for damages to their salvage project on the wreck. Frick and Gasque had been negotiating with financiers for books, films, and even computer games based on their expedition. They asserted the INA maligned their ability to handle the job professionally. All three groups, Nomad, Caribbean Ventures, and the INA were claiming contractual agreements with the government. This was going to be good.

Nomad had close to twenty people involved in the operation; Dr. Nancy Desautels, an archeologist from Santa Ana, California,

Roger Miklos and his girlfriend   Photo: Jim Ryan

Roger Miklos, his son, his girlfriend, the cook, the engineer, a Turks and Caicos police officer, two Dominican crew members who came with the boat originally, several investors, Captain Jon, Jimmy, and me. The cook's name was Brooks and he was a cowboy cook straight off the western range. As we finished the fine meal which Brooks had prepared that evening, we were increasingly aware of a tropical disturbance miles away but close enough to send large swells and strong winds our way.

The winds and the swells picked up quickly. It was too dangerous to transfer Jimmy and me to our berths on the *Capt. Jack*. The storm must have been traveling quite rapidly because it was upon us in no time. The seas increased from light to moderate, to 20-foot plus. It was impossible to stand; it scared the investors. Some ladies were crying. Captain Jon, Jimmy, and I stood watch in the wheelhouse with a view of the *Capt. Jack* tethered to the *Heather Glynn* bucking the seas. We could barely see her in the wind-driven rain as we would throw the spotlight on her hoping she was hanging on in one piece. The tops of the seas were now blowing over the wheelhouse of the *Heather Glynn*. I wedged myself between the steering wheel stanchion and the bulkhead and hung on. The storm really rocked us for a couple of hours and then ended almost as fast as it had started. When the sun rose a few hours later the dinghy with my two dive bags of gear had sunk. Jimmy and I dove overboard to retrieve what we could from the bottom, finding my dive gear bag lying under the dinghy but there was no trace of my second gear bag. The missing bag had contained my regulators, my gauges, the high-dollar stuff, wrapped up in my wetsuit.

We re-floated the dinghy and mounted the motor to it, which fortunately had been removed before the storm. The dinghy was used to pull me back and forth around the area, hoping I could recover my missing bag and gear. It took about 20 minutes before I spotted my bag laying on the bottom. It had been nearly neutral in buoyancy with the wet suits inside and zipped up it also had held a little air. Just enough air that it bounced along the bottom until a hole was ripped in it causing it to settle to the bottom. I breathed a sigh of relief that I had my gear. In less than 24 hours, I had been pummeled by a tropical storm, lost my gear, and had recovered same. I was having fun now!

Nomad's targeted wreck was near Molasses Reef on the south-western edge of the Caicos Bank near French Cay. We set sail for the Caicos Bank traveling slowly through the shallow waters enjoying the view. We also had picked up a young local islander along the way to run errands for us in his little skiff. We had a SCUBA tow, which was basically a large plastic wing with hand grips you would hang on to while being pulled behind a boat. It worked well... tilt it up and up you would go. Tilt it down and down you would go, and you could even go sideways with some experience. Since Captain Jon had to be aboard the *Heather Glynn* all the time, Jimmy was acting as the captain of the *Capt. Jack*, and he could not leave her. That left me. The young islander would pull me along at a moderate speed, but not so fast that I couldn't hold on. Day after day he pulled me behind the boat, above the reefs, over reef after reef. The visibility was a hundred feet plus horizontal, just about everywhere. This was good, and this was bad. I had company much of my time cruising along, and I could see them off to my side and behind me. They had no problem keeping up with me. Swimming alongside me were sharks. I saw a variety, black and white tip sharks mostly, with a Tiger or a Hammerhead shark thrown in occasionally. The bottom was strewn with ship-wrecks and shipwreck material. Ballast stones littered the ocean bottom along with brass ship pins, spikes, anchors, sheathing, and all types of shipwreck-associated material. Stopping to investigate these sites made for slow going. We couldn't go a mile without finding an-other shipwreck. There were some modern wrecks also, steel hulls rusting away and blending into the bottom. I could tell Jimmy wanted to switch places with me so badly... but it was all mine to discover.

When Captain Jon ended the daily freshwater showers, the investors went home. The investors had experienced all they wanted to experience. The romance of treasure hunting and life at sea soon lost its appeal. The days were gorgeous, the weather nice and the wa-ter clear. Days passed into weeks, then into a month. Every day I was being pulled like a fishing lure through the azure-colored, crystal clear waters, finding shipwreck after shipwreck. But we hadn't found the correct one yet.

We spent our nights sometimes listening to the single-side-band radio, which can be quite entertaining. We heard the voices of the smugglers, sailors, ham operators, the ship to shore calls. Then other times we listened to the local island am stations which can be hilarious. The food was wonderful; Brooks, our cowboy cook, really knew what he was doing in the galley. When he wasn't cooking, he spent his spare time throwing a lasso at one of the towing posts on the fantail. In the evenings, when both boats were at anchor, the rest of the crew would jump off the sides and go for a swim or snorkel around a little. Just out of orneriness and boredom, we would wait to flush the head until an unsuspecting swimmer was in proximity, then we would bomb away. When a piece of toilet paper would wrap around a person's snorkel, we would howl like children.

While we searched for the wreck Roger was calling the *Pinta*, the wreck was getting attention from two other groups. Key West treasure hunters Olin Frick and John Gasque, who claimed to have discovered the wreck in 1977 had partnered with Dallas millionaire William R. Reilly, and the two treasure hunters with a crew of 23 divers and archaeologists had planned to return to The Turks and Caicos. Once they heard Nomad was in the area, the name-calling and threats began. Also plotting in the shadows was the Institute of Nautical Archaeology at Texas A&M University. This group most likely heard of Frick's and Gasque's discovery through the national news channels. They contacted the British government, bypassing the local government authorities, expressing their concern for proper excavation. They made no mention that Nomad had an archaeologist with a doctorate on board the entire time. The two groups hadn't communicated with the Nomad group, preferring instead to fighting it out in the press, slandering Nomad, and threatening to "take action", whatever that meant. At the time we didn't know that there was a different type of storm brewing far away from our tranquil seas.

We ran into Providenciales every couple of weeks to get supplies and to use the phones. I went along once with Jimmy on the *Capt. Jack*. Captain Jon had to stay onboard the *Heather Glynn*. The crew would usually give Captain Jon money to buy some El Presidente, good Dominican Republic cervaza. This time they placed me in charge of securing the beer for the crew. Beer never seemed to have lasted very long onboard the two ships. It would be consumed in a couple of days. You couldn't keep your beer in the fridge, it would mysteriously disappear. I would stash mine under the gear in the forward anchor locker. "Warm beer is better than no beer", Confucius say. This time when we pulled into the small harbor, we docked next to a sailboat. The owners of the sailboat were ashore getting supplies and had left their small 8-year-old son aboard to watch over the boat. I also stayed aboard our boat while Jimmy ran into town for supplies and my beer order for the crew. This little guy had spent his entire life on this little sailboat with his parents. His parents were of different nationalities, so he was bilingual. He spoke French, Spanish, and English, fluently. He spoke like an adult since he had never really been around other kids. He offered me a tropical cocktail, which I took him up on. It was delicious and potent. We spent several hours that day talking, and he had a few questions for me. He had flown to Miami with his father one time to pick up a part for their boat. While in Miami, his father had left him in the hotel room watching TV and he'd watched cartoons much of the afternoon. He asked me several questions about Batman and Superman. How could they do this and that? Why could they fly and do other superhero things? He had never seen a cartoon before that, and the borderline between fiction and nonfiction was confusing to him. He had no need in his life for any type of fantasy — his life was already full of wonder and adventure. It bothered him, he couldn't understand what was real and what wasn't on the TV. I tried my best to clarify it for him, that it was make-believe for fun.

While we talked, a few locals walked by asking for a little piece of gold, just a little piece. It was pretty obvious that Nomad had made a big impression on the native population. I still remember my time spent with that little boy that afternoon, I think of him often. I wonder what became of him, and what and where he might be today. Jimmy returned, and we headed back to anchor up with the *Heather Glynn*. Once back aboard the *Heather Glynn* I distributed the beer among the crew. They grabbed it and were about to tilt a few bottles when I said, "Don't forget your change." They all stopped in their tracks, "What change?" It seems this was the first time they ever got change back. Captain Jon would not be happy with me. The beer tax had been exposed.

The author aboard the Capt. Jack in the waters of Providenciales
Photo: Jim Ryan

Chapter 16
The Find & Fleeing
The Confusion

I never tired of being towed behind the skiff, day after day. The sharks that followed me were now a common sight, and my eyes were quick to spot any irregularities on the bottom. This one particular day I was being towed over a reef, which extended skyward suddenly from deep water. I almost collided with the reef once or twice. Then suddenly, I looked over to my right, and let go of the rope. Below me, I was almost surrounded on all sides by a small ballast pile. On top of the ballast pile were two bombardeta (early Spanish cannon), two or three versos (small deck guns), and an anchor[1]. I swam down and examined the ancient wreck. It looked untouched. The ballast was not loose, but stuck together, almost glued together by marine growth. Many more of the small versos were scattered across the bottom just to one side of the wreck. I could see cannonballs on the ballast pile, some were iron and some stone. It was a beautiful sight to see! This was the oldest European shipwreck found in the Western Hemisphere. For over 450 years, the wreck sat on the bottom surrounded by the reef, apparently undisturbed after its loss. I was in awe. I broke the surface and announced that our search was over, she was below me.

We anchored the two boats maybe a quarter-mile away from the reef and the wreck site. I boarded the *Heather Glynn* and shared my observations with Dr. Nancy Desautels, the archaeologist, and

---

1   A verso is a smaller cannon, with a handle and a dual-tang mount that could be affixed to the ship's railing. It was designed to fire smaller projectiles and could be operated by one man.

we all agreed this was the wreck. Now, here is the odd thing; Roger didn't dive, and neither did the archaeologist. So both of them relied on what the divers were reporting. Those divers were Jimmy, Roger's son Darrell, and me. I'll swear in any court of law in any country, we never removed one item from that wreck, not even a cannonball... nothing. Everyone was aware of the importance of that wreck so there was no plan to disturb it. We did as we were instructed; we notified

A "verso" annealed to the seafloor
Photo: Jim Ryan

The Turks and Caicos government, and they told us to stay on the site to protect it. We sent a boat over to Provo and picked up a Turks and Caicos police officer to serve as an observer. We spent the next few days photographing and mapping the wreck site. We had also located another large anchor a short distance away, but we didn't feel it was associated with the ancient site. We waited for the island authorities to tell us what to do next.

The day was a little breezy. I could make out a small boat headed our direction, occupied by maybe three or four individuals. They eventually anchored near the wreck site. We attempted to contact the unidentified boat by marine radio but got no reply. I alerted Captain Jon, Roger, and the rest of the crew, including the police officer that an unidentified vessel was close to the wreck site. Captain Jon, Roger, and the police officer readied our skiff to confront the occupants of the intruder. We had arms on board both of our own vessels. You would be crazy not to have protection in these waters. I remember handing a rifle to the police officer to take with them as a show of authority. The police officers in the islands carry guns, but they are small pistols in a little purse-like holster, so the rifle would be more obvious.

We didn't know who these interlopers were. Captain Jon, Roger, and the police officer motored out to confront them. We found out they were representatives from the Institute for Nautical Archaeology and their spokesperson was Donald H. Keith. The conversation was terse, with each party identifying who they were and what authority the government had given to each. There were no threats of harm as the police officer identified himself, presented his credentials, and explained his purpose for being there. Donald Keith didn't want to discuss the issues or come on board for a cup of coffee that day, which he was welcome to do. He claims to have come onboard the *Heather Glynn* but that is a lie. He also states that he saw an anchor and artifacts in a box on board, and that's a partial lie. He never came on board. We had recovered a large anchor, which was an isolated find, not associated with any known wreckage. He asserted that these artifacts were somehow associated with the *Pinta* site, which is simply not true. It seems this was Donald's big chance to be a hero in the media and be acknowledged by his peers as he portrayed himself fighting the treasure hunters who were "plundering and looting historic shipwrecks." Donald and the INA didn't want to fight in a court, nor did they wish to acknowledge the island's government authorities. They had gone straight to the British government, superseding The Turks and Caicos government[2].

Donald H. Keith didn't waste a minute getting back to dry land and a telephone. The headline in The Washington Post read, "Armed Divers May Have Looted Shipwreck Claimed to Be Pinta". The article, which immediately ran worldwide, stated: "An unidentified group of salvagers reportedly were anchored over the shipwreck site this week and may have escaped with several artifacts, according to officials of The Turks and Caicos Islands, which has jurisdiction over the shipwreck." We didn't know any of this then; we had no idea that Donald Keith had fabricated such an enormous lie. We had over a dozen people

---

2   The Turks and Caicos are a protectorate of the British Commonwealth, and their authority in international matters is superseded by that of the British Government.

The anchor was raised to the surface and landed aboard the Capt. Jack.
Photo: Jim Ryan

on board who could confirm it was Donald Keith's desire for attention coupled with dirty politics that were the actual crime story that day. Donald has retold the tale of the "armed pirates" over and over through the years... the incident must have been the highlight of his career. Poor fellow. This would be the closest he'd ever get to being "Indiana Jones".

It wasn't long before the government officials who Roger had an agreement with invited the Nomad crew to Grand Turk. Donald Keith's false accusations had reached ears on both sides of the Atlantic; the local government and the British government. The local authorities were now being called into question for their agreement with Nomad Treasure Seekers. When we arrived at Grand Turk, we refueled and took on fresh water. Captain Jon, Roger, and Dr. Desautels went ashore and met with officials from the island government. The news they came back with was not good. An investigation was underway, and we were placed under house arrest and detained. Both vessels and all the crew were under house arrest. No one was going any-

The Capt. Jack laying along side of the Heather Glynn is seen here with the author (back to camera) and Jon Christiansen on deck. The fellow in the background is the son of one of the project investors. Photo: Jim Ryan

Left to right; Jim Ryan, Randy Lathrop, and Jon Christiansen pose with an ancient anchor, one of many in the area, alleged by Donald Keith to have been removed from the site of the "Pinta". Neither allegation was true as this particular anchor did not come from the erstwhile "Pinta" wrecksite, and the cited wreck was not the "Pinta".

137

One of the "bombardeta" cannons found grown fast to the ocean's floor. These ring-bound iron guns were used during the 1500s
Photo: Jim Ryan

where. They also brought back two new police officers, one stationed on the *Capt. Jack* and the other on the *Heather Glynn*, 24-hours a day. After a month of investigating over twenty shipwrecks while looking for the *Pinta*, I had brought topside maybe a bucket full of brass ship pins and assorted spikes, some square ballast, some iron items, and all were junk from other scattered wrecks. We had no idea where this was going, so Roger instructed us to throw it all overboard. I didn't feel this was necessary, as it was just scatter from many wrecks, some of them modern. So while the INA continued to prosecute us in the media with no evidence at all, we sat there, waiting, and waiting.

With each passing day, we became more concerned for our fate and that of our two vessels. It was obvious no one wanted to hear our side of the story. Donald Keith was literally building a career from this conundrum, as you shall see. Our two boats would make a hell of an addition to The Turks and Caicos navy. They could start their own defense force, which they did much later. It was obvious the British Commonwealth governors had called their colonial dependents on the carpet for their handling of this matter. We were stuck in the middle of a big mess. We needed a plan.

The Turks and Caicos police officers were friendly guys. They would run two shifts for our house arrest, a day shift and a night shift. We would use our Zodiac inflatable to pick up the day shift in the morning as we returned the night shift to shore. We picked up the night shift when we ran the day shift back to shore. You get it. The plan was as follows. We would pick an evening and when we dropped the day shift off, the boats would slowly weigh anchor, holding their position, so as not to be noticed. Before the night crew could get on board the Zodiac, it was to take off and join the other two boats now underway and escape. I was to be the guy in the Zodiac... I love an adventure! Our concern over the possible seizure of our vessels had not abated. It was time to make a move.

When the fateful day arrived, we shared our plan with the rest of the crew. The shift change came around just as the sun was setting on the horizon. I had both police officers from the day shift in the Zodiac and headed to the long dock to drop them off as usual. The two police officers who were to relieve them were way down at the end of the dock near the shoreline. The day shift guys started the long walk down the dock to greet the other two. They met and started talking. This was my chance. I cleared the lines and headed out as fast as the Zodiac would take me. No one had noticed I had slipped away yet but not for long! A native was fishing off the end of the dock. He was using a soda can wrapped with fishing line like many folks there do. When I zipped past him I snagged his fishing line, and he began to yell, "Hey Mon, my line Mon, my line!" The cops at the end of the dock now looked my way and saw me trying to catch up with the *Capt. Jack* and the *Heather Glynn*. Both had weighed anchor and were making good speed due east out into the open Atlantic ocean. I was waiting for gunshots to ring out and bullets to fly by... I was sure of it. But that never happened. I came up behind the *Capt. Jack*, which was the slower of the pair, and threw Jimmy a line, which he made fast as we continued our escape. I stayed in the Zodiac for several miles until we felt it was safe to slow down so I could rejoin my crew aboard the *Capt. Jack*. Once I was safely on deck, we secured the Zodiac and headed out to the open sea in the descending darkness.

Everyone breathed a sigh of relief once we were on our way. But we still had several issues to deal with. We had not cleared immigration or customs with The Turks and Caicos Islands on our departure, which would impact us upon our return to the United States. Nomad officials had established communication with stateside authorities and were talking to Senator Alan Cranston of California asking for help. We'd evidently made more headlines with our unauthorized departure.

We sailed through calm water on our trip back to the United States. At night on the *Capt. Jack*, to break the boredom we would take Cyalume light sticks and place them in a plastic jug and use it for target practice with a Mini-14 carbine as we pulled it behind the boat. This was the same rifle that scared Donald Keith and worked so well in his "bad treasure hunters" narrative. What a joke. I hope he reads this someday.

When we arrived at Fort Pierce in the evening, Customs and Immigration had closed for the day and we were to clear the following morning. Of course, as soon as we tied up we had visitors as friends and acquaintances showed up to greet us and hear of our adventures. We cleared Customs and Immigration the next day with the help of Senator Cranston. We were all very grateful. None of us had gotten paid and to Roger's credit he and Captain Jon drove down to the Florida Keys to sell some of Roger's coins to a collector. Roger paid us from the proceeds.

The *Pinta* story continued gaining steam in the press, the bulk of it coming from the lips of Donald Keith who continued to trumpet his Hollywood version of events. So Captain Jon and Roger started talking to the media themselves presenting our version of the story.

The story was headline news in many major newspapers across the country. I recall the Miami Herald headline read, "Unidentified Pirates of Pinta Identified". Several people knew I had been on a project in The Turks and Caicos, and frequently people would ask, "Hey, that wasn't you down there on the *Pinta* was it?" Florida Today did an article titled "Shipwrecks Lure Divers into the Deep". It didn't say into deep "what", but I had more than one answer for that.

The expedition vessels underway with the Capt. Jack astern of the Heather Glynn

Chapter 17
Aquatic Adventures

After the Nomad Treasure Seekers venture, I went back to work for Aqua-Tech on several jobs. One job we did up in the panhandle required that we build a revetment of cement bags along the shore where an underwater cable was exposed. This required the loading of 50-pound bags of cement, thrown in one of our skiffs and then taken to where the cable ended on the bank at the wooded shoreline. We had to make several trips in the skiff, loading and unloading the brown bags from the skiff to the shoreline. A mile or two away in a high-rise condo sat a person watching us unload these brown bags onto the wooded shoreline. "Miami Vice" was hot on the airwaves at the time and these "brown bags" being cast ashore just didn't look right to this guy in the high-rise. We were close to unloading our last load of cement bags, coasting to a stop with our load. Suddenly we hear a rustling in the woods and hear a voice shouting for us to "Freeze!" Out of the woods jump several sheriff deputies with their guns drawn. Fortunately, one kick at a brown bag revealed its contents to be cement, not the devil's lettuce they suspected. They had received reports that we were off-loading bales of the evil weed. Everyone had a good chuckle, and the deputies apologized for the misunderstanding.

The Aqua-Tech job was part-time whenever they needed an extra hand. So I was happy when I got a call out of the blue from a guy named Walter. Walter Hornberger was a sales rep for US Divers Company, he also had his own businesses on the side. One business he owned provided life-guards for public and private swimming pools.

Walter would use young guys to work for him. He made the assignments, wrote the contracts, and handled all the billing. He had just started a new business somewhat along the same lines. The difference being that he was providing diving instructors to cruise lines for their aquatic programs. He called his new company "Aquatic Adventures". He had just landed a contract with a new cruise line, Scandinavian World Cruises. They had ships in Miami and Port Canaveral, sailing to The Bahamas for one-day cruises. Walter, being the sales rep for US Divers, asked about instructors when he would visit dive shops on US Divers business list. One shop he knew well was the 76 Fish and Dive Center in Cape Canaveral, owned and operated by Jerry Fish and his family. Jerry gave Walter my name as someone who might be interested in a job. I met Walter and his wife at the Howard Johnson on US 1 in Cocoa (it's been long gone since). Walter explained that he had contracts for two Scandinavian ships. Both ships would sail to Freeport, Grand Bahama for the day. The *Scandinavian Sea* sailed from Port Canaveral, and the *Scandinavian Sun*, sailed from Miami. He needed certified dive instructors to run snorkel trips as a shore excursion once the ships docked in Freeport. His business required the dive instructor status because only certified instructors could carry insurance. Walter asked if I was interested and if I knew of any other instructors. I told him I was interested, and I knew of another instructor. Jimmy just had gotten his instructor certification and when I asked him he was game too. The first ship to set sail would be the *Scandinavian Sun* running daily trips to Freeport. The ship would leave Miami at approximately 8:30 AM and would dock in Freeport around 11:00 AM. Passengers would go ashore for the day and needed to be back by 6:00 PM to depart for Miami where they would disembark at 10:00 PM.

The *Scandinavian Sun*, built in 1968, was 441 feet long, with a beam of 71 feet, and she weighed 10,448 gross tons. She had just sailed from Hamburg after being remodeled. Much of the ship was still being worked on when Jimmy and I stepped on board for the first time. This was a big change for both of us, being normally accustomed to smelling diesel fuel fumes all day and a wet suit full of fine sand. This job came with sailor uniforms with a petty officer designation,

There was no shortage of attractive girls in the cruise business.

and several crew cafeterias to choose from for our meals. Our job was to sell the snorkeling program at the shore excursion office alongside the rest of the shore excursions being offered. When the ship docked in Freeport, we would gather up our customers, load them on a bus, and take them to the beach where there was a small reef just offshore. When we were not selling the program or working our tours, we worked as cruise staff on the ship in the evenings. Cruise staff performed a variety of functions on board. They would act as the MC (master of ceremonies) for shows, run the spotlights, run the bingo games, and assist with many of the ship's evening activities. Our only issue was not having assigned cabins; we had one cabin assigned to four diving instructors. We used the one cabin for our luggage storage. Once the passengers disembarked for the evening, we could have any room we wanted until the next day. Every day we made the trip to Freeport and back to Miami. And the girls... there were lots of pretty girls, all kinds of girls. There was no shortage of girls, dancing girls, showgirls, big band singers, musicians, and entertainers mixed in with the crew. There were comedians, rock-and-roll bands, ventriloquists, lounge singers, etc. Everyone meeting one another for the first time and starting on a new ship. It was exciting!

This sure looked like it was going to beat commercial diving, dirty barges, and long hours.

I forgot to mention Walter and his partner Bill Myer were both retired New York State Troopers. Bill would be our direct supervisor and would sail with us occasionally. Walter had contracted for the production of an underwater slide show by a well-known photographer in the Keys. The slide show presentation was about 20 minutes long. We were provided with a script that was to be read as the narrative during the slide show. Every one of the instructors hated it. It was difficult to stay with the pre-programed slides during the talk. We had to do this in front of a couple of hundred people, and it mortified several of the guys to speak in front of a large crowd. All the instructors would gather at the back and watch the poor soul selected to talk, who would usually crash and burn. It often was painful for all parties involved, between laughs. I practiced and edited the script so it wasn't so awfully boring, and I became comfortable giving the spiel. The other instructors would do anything or give me anything if I did the talk for them. So the entire time we were required to give the presentation, I drank for free, courtesy of my bashful co-workers.

We quickly made friends with the cruise staff, cruise director, a few officers, entertainers, and musicians. This didn't seem like work at all. After hours we all hung together drinking and just having a good time. Our snorkeling beach was several miles from where the ship docked, so we took an old school bus to the beach. An old Bahamian man always drove the bus to the beach, and he had a microphone. During the drive, he would bore everyone with his regular boring talk, most likely picked up from other boring bus drivers in the islands. "Welcome to the islands, there are over 700 islands in the..." blah, blah, blah. It never got much better. We endured it on every trip until one day when the famous comedian and ventriloquist, Bob Rumba, asked the driver if he could talk and the driver handed the mic to Bob. We knew this was going to be interesting. As we passed a few brightly colored Bahamian houses, Bob started his talk. Bob talked about how the Sesame Street Gang had come to The Bahamas to teach the people the different primary colors, and that's why the houses were the color they were. As we passed a corner where three

The author, left, aboard the Scandanavian Sun with an unidentified fellow crew member.

Bahamian hotel maids stood on the sidewalk, Bob yells out, "Ladies and gentlemen, the Supremes on vacation here in The Bahamas." He talked about the famous skinny Bahamian pine trees as we passed by the pine trees. Everyone on the bus was in stitches, everyone except the driver who eventually snatched his mic back when Bob couldn't keep from laughing either. I'll always remember that bus trip and the look on the face of that poor, unknowing driver when he knew he had made a mistake. Our snorkel site was just off the beach, the passengers would wade and then snorkel to the little reef from the beach. There was a clubhouse on site with refreshments. As far as gigs go, it didn't get any easier than this one.

We returned to the ship around 4 o'clock, the passengers would hit the bar and the buffet and we'd be back in Miami at 9 or 10. Once the passengers had disembarked, the city of Miami was all ours. The musicians, dancers, and divers would gather and hit the clubs. One of our favorite places was "Tobacco Road" at 626 South Miami Avenue, attracting everyone from some of the Magic City's most notorious gangsters, including Al Capone, to war vets, drifters, and party animals alike. It was a long skinny bar with clouds of cigarette smoke, and many famous musicians. The bar closed for an hour at 5 AM to sweep the floor. Another place called the "Eighteen Hundred Club", served a free breakfast starting at 4 AM. It was an after-hours club. It cost $25 to join and closed at 6 AM. They had a piano bar and several rooms. Many times we would come stumbling back to the ship at 4 or 5 AM as the quartermaster would look the other way as we staggered up the gangway. If we were hungry, we often would sneak into one of the ship's many galleys and raid the refrigerators. Dancers and divers giggling and rummaging through everything, hoping to find a treat, maybe some lox and bagels. Then at 8 o'clock, passengers would board and it would start all over again.

The weather could make it interesting. The ship set sail regardless of the weather, every day. When we would cross through the Gulf Stream, we all knew. The swells and waves would increase and the stream would toss and roll the big ship from side to side, then from bow to stern. The ship's rolling and reeling was in slow motion, which made many of the passengers sick. When the weather was rough, it wasn't a pretty sight. Seasick bags lined the rails, all along the insides of the ship. We were forced to cancel our program when the weather got bad. Instead, we would walk to Pinders Point, close to the ship terminal in Freeport. We had a favorite bar called "Fat Mans", a native bar we would take over on rainy, windy afternoons. We had some good times at "Fat Mans". Many years later, when crack cocaine invaded the islands, a crackhead shot and killed Fat Man.

Once they got the ship at Port Canaveral ready, Jimmy went back to Cocoa Beach to work onboard the *Scandinavian Sea*. He could sleep in his own bed at night, and besides that, working the cruise ships didn't really seem to be up Jimmy's alley. The cruise line was having issues keeping our snorkeling location contracted, and eventually, they lost the location. When they lost access to the snorkeling site, Aquatic Adventures kept two dive instructors on board the ship to satisfy the contract. It wasn't Aquatic Adventures' fault the cruise line lost access to the site. What this meant was that Steve, the other dive instructor, and I were to stay on board until further notice. We were being paid to cruise. The ship wasn't happy about it and

147

wouldn't assign us any further cruise staff duty to help break up the time for us. They wanted to see us leave so they could get out of the contract. We slept, we drank, we ate, we waited. Our shipmates were on our case all the time about how easy we had it, but we hated it after a few months. During this time I met my soon-to-be-wife on board. She was the shore excursion manager. It was a whirlwind romance... that's what cruise ships do to people.

One early morning as we were docked in Miami, the captain called me up to the bridge. I did not know why the captain would summon me to the bridge. Once on the bridge, the captain shared with me that a heaving line had become wrapped up in one of the ship's propellers. A heaving line is a smaller line they tie to the large lines, the heaving line has what is called, "a monkey's fist". It's a large knot at the end that lends it some weight so it can be thrown from shore to the ship, and vice versa. He asked if I would dive under the ship and free the line from the prop. "Of course, I would," was my reply. I then asked him to be sure that prop didn't turn while I was down under the ship. Actually, I inquired about the "keys" to the ship and he had a good laugh about that one. I was soon crawling around under the massive ship. It can be very disorientating crawling around like a fly on a ceiling, the only difference is your upside down, crawling on the bottom of the ship. I located the two large propellers and found the offending line wrapped around and around one prop. I sat on top of the prop and cut away at the line, and it took me close to an hour to cut it off. I cleared the line just in time for an on-time departure to Freeport. The captain wrote me a letter of commendation thanking me for my service to the ship in a time of need. He also asked what he could do for me. I asked if it would be possible to open the disco for a staff party after we disembarked passengers one evening. He said no problem. The disco was opened the following night with free drinks. It was a good time, all thanks to a poor throw of a monkey's fist.

I had received a message on my answering service at home from Rex Stocker. He asked that I call him when I could. I mustered some more quarters and called Rex. Rex had partnered with John W. Mecom, Jr., an independent oilman in Houston, Texas, who also

owned the New Orleans Saints football team. He owned and operated the John W. Mecom Company, which is primarily involved in
real estate investments and oil/gas exploration and production. John
was also interested in the treasure salvage business and had dedicated a 96-foot crew boat with three 1271 Detroit diesels called the
Rio Grande to the enterprise. They decked her out with all the latest
electronics; weather fax, SatNav, blowers, and an osmosis water system. The boat was doing some remote sensing survey work in The
Bahamas south of Cat Cay. It had been slow going down there, and
Rex didn't understand exactly why that was the case. Rex asked if
I would be interested in joining the crew. Maybe I could help him
understand what was going on. The boat had been sailing out of Ft.
Lauderdale, back and forth to The Bahamas. I had enough of being
a professional cruiser, so I agreed to join. When I spoke to Walter at
Aquatic Adventures and told him that I was leaving, he shared that he
was close to an agreement with the cruise line and our time on board
was about over, anyway. I would be just up the road from the ship in
Miami, so it was a happy ending.

Chapter 18
The Rio Grande

I boarded the *Rio Grande* in Fort Lauderdale at the Pier 66 marina. The marina was home to traveling million-dollar yachts and a few fishing charters lined up along the A1A road frontage. Waiting for me on board was Sam Staples, who was acting as operations manager. Sam was from the Melbourne Beach area and had been flirting with the treasure business for some time before partnering with Rex on the *Rio Grande*. The skipper, Captain Bob, was from Rhode Island, of Portuguese descent, and this was his first command. Ian Kelso, the son of Doc Kelso who was involved in the original Real Eight group, was a crew member as well. Ian, a former Navy Corpsman, had a hard look but a kind heart. Bobby Heater, also from Melbourne, Florida, was another crew member. He was a friends with Ian Kelso and Rex Stocker. Completing the crew was a former Navy Seal, Mark Wolf, from Cape Canaveral. Mark viewed me with suspicion, and I viewed him the same initially. But we would become good friends in the months to come. We had a cook on board who would soon rotate out because of an affinity for alcohol. That was too bad because the man could cook up a storm. He was fine as long as the boat was at sea, but once he hit the shore, he became a staggering drunk. I was to bunk in the same cabin as Captain Bob, which would prove to be a mission in tolerance for me. There is no other way to say it, but Captain Bob had a big mouth and an abrasive nature.

The first several weeks we spent doing routine maintenance and preparing for an extended voyage south of Cat Cay. In the evenings I would drive to Miami and stay the night onboard the *Scandinavian*

Sun with Penney, my new girlfriend. Penney was from Louisiana and had the accent to prove it. She had been working on cruise ships for a couple of years; she worked in the Miami offices for NCL and then Scandinavian Cruise lines where we'd met. She had dark curly hair, was thin, with very fine features, fair-skinned, bright eyes, and a winning smile. She had an extended family back in Louisiana where she could trace her French roots back to the 1600s. She had lived in Miami Beach while working in the NCL offices, but jumped at a chance to work onboard when offered a job by Scandinavian Cruises.

We departed Fort Lauderdale after picking up a new cook, a wanna-be biker from Fort Lauderdale, who turned out to be a terrible cook. When Mark ordered his eggs over easy the first morning, we heard the cook say, "Whoops," when he broke the yolk. Mark said, "Shit, we're screwed."

We sailed for Cat Cay, and because John Mecom was who he was, they allowed us to dock at the exclusive private island for millionaires. They allowed us to go ashore at Cat Cay and drink at the native bar for the workers. But we were to be on our best behavior at all

The Rio Grande quay-side in The Bahamas

times. The following day we started our magnetometer survey work south of Cat Cay, past Ocean Reef. We were pulling a proton mag following an imaginary grid, and we were going fast. I was surprised at the speed they were pulling the mag, at around 6 to 8 knots. Sam had his own ideas, and nothing could sway him from them. We would spend several days magging, and then we would go back to Cat Cay for fuel and dock for the night.

Captain Bob could be a real jerk and I was over him and his big mouth. One night after having a few drinks at the native bar, I'm walking down the golf cart path back to the boat when I hear laughing, and somebody coming on a golf cart. I jumped in the bushes, just as Captain Bob and Mark come flying around the corner in a stolen golf cart, drunker than a couple of skunks. I followed them in the dark far enough just in time to see them wreck the golf cart and run away. I walked back to the boat with smug satisfaction, having seen the entire episode. The next morning as I lay in my bunk, Captain Bob starts his morning routine. "Drop your cock and grab your socks," he would yell every morning.

"Kiss my ass," I replied. I stunned Captain Bob.

"You better watch your mouth," he said.

"Or what?" Before he could answer, I told him how I had seen him wreck the golf cart last night and he needed to treat me better. This got his attention. The management of the island knew about the cart, but they didn't know who stole it. "So if I was you, I wouldn't start any shit," I advised him. That seemed to work like a charm for the moment.

We continued to mag at high speed, much to my consternation. I suspected that our magnetometer could not register anomalies while running so fast[1].

One sunny afternoon, as we were doing our survey work, we spied a large brown object in our path. As we came closer, we could see it was a large bale of pot. The crew groaned as we passed by the bale, for various reasons. In other potential scenarios, it might have

---

1   Magnetometers use a burst of power to detect anomalies, and must recycle themselves between bursts, therefore it is necessary to tow them slowly to guarantee their effectiveness.

been considered a windfall — or a ticket straight to federal prison. We continued to pass the bale repeatedly, and the reaction was nearly the same every time. Finally, Sam got out one of the Ruger Mini-14's. We had one for every crew member. They had banana clips and were rigged to shoot full auto with a piece of fishing line tying the trigger to the bolt. The next time we passed, Sam shot the bale full of holes, which also tore it apart. The weed drifted off into the sunset along with our dreams of illicit booty. No booty for these pirates.

Some days we would stop and shoot a few fish or gather some conch for conch salad, which most of us loved. We had gathered and cleaned a nice pile of conch one afternoon and gave it to the biker cook to make conch salad. The biker took the conch and the next time we see it it's cut into medium size pieces, with lettuce, tomatoes, and Green Goddess salad dressing. When we saw what he had done with our conch, our hearts just sank. He did not know what conch salad was or what was in it. Conch salad, for those who don't know, is more ceviche than salad. It's finely minced conch pieces, with minced onion and peppers, small minced tomatoes, covered with lemon or lime juice to cook in the acids. We should have known better; the biker cook was still wearing his boots and bandana on the boat and we should have guessed he wouldn't know what conch salad was or how to make it.

Sam got heartburn easily, probably because of all the cigarettes he smoked, but he would blame it on the food. The cook would put baking soda in his dishes because he was afraid of being accused of giving Sam heartburn. We wouldn't stay at sea long because it seemed Sam always had a problem that required returning to Fort Lauderdale to fix it. Sometimes it was something silly, like the saltwater converters on the boat. I could see now why things were going as they were.

There is one particular incident that could have turned deadly for all of us. The area we were magging was a remote spot south of Ocean Cay. We knew the area to be active with drug smugglers, so routinely we stood watch all night. We would each do 4-hour shifts in the wheelhouse watching the radar. We would set up a 3-mile perimeter alarm on the scope, and if that was violated, we reduced it to 2-miles, and then when the blip on the radar got within a mile,

The Rio Grande was used by John Mecom as an excavation vessel, searching for treasure lost on the 1715 Fleet. Here it is seen working in the area of the Cabin Wreck (See Page 4).

we would wake everyone and have them take stations. Every crew member had been issued a Mini-14 rifle, with banana clips of .223 ammo taped back to back. The guns were rigged with fishing line so when the bolt was pulled and let go it would act as an automatic and empty the clip.

This night Bobby Heater was on watch. I couldn't sleep and joined Bobby in the wheelhouse. There was a small group of rocky little islands near our anchorage. It was very dark that night. Bobby said, "Hey Randy, it looks like that island is moving." As I watched the radar screen, I saw one of the small islands move. It stopped for a moment, then slowly began to move in our direction. It was going slowly, but it penetrated the three-mile perimeter. Bobby knocked the scope settings down to 2 miles, the blip slowly continued ambling once more until it was now inside the 2-mile radar ring. Whoever it was, they were coming to pay us a visit. When the blip broke through the 1-mile radar ring, we sounded the alarm. Captain Bob immedi-

154

ately came to the wheelhouse, while the others armed themselves and took positions along the rail facing the oncoming blip.

Captain Bob dialed up channel 16 on the marine radio and hailed the unidentified vessel: "Unidentified vessel, this is the *Rio Grande*, identify yourself." No answer. The blip kept coming towards us. Captain Bob repeated, "Unidentified vessel, this is the *Rio Grande*, identify yourself." We turned on our searchlight and scanned the dark horizon for any movement. Captain Bob was anxious now and repeated himself with more authority this time, "Unidentified vessel, this is the *Rio Grande*, identify yourself." No answer. They continued towards us. Captain Bob now told everyone to get ready but we were not sure for what exactly. As Captain Bob once again ordered the vessel to identify itself, there was nothing but silence on channel 16. Eventually, I could make out a white hull in the darkness. This was not good, I did not know where this was going. As the vessel got closer, I could make out a white hull and a RED STRIPE! Suddenly the radio came alive with, "This is the United States Coast Guard." We were all relieved and pissed off. We'd come very close to firing on them in the dark and that would have been the end of us. Captain Bob did not hold back and chastised them for their actions. They came alongside briefly and were soon on their way[2].

I kept in touch with my girlfriend Penney on the *Scandinavian Sun* via a single-sideband radio. We would have arranged times to call one another ship-to-ship. The radio officer on board the *Scandinavian Sun* was our friend, so it was no problem.

The *Rio Grande* continued to make the trip between Florida and The Bahamas for several more months. On my last trip aboard the *Rio Grande* we were coming back to Fort Lauderdale at a good clip. The cook was in the wheelhouse looking aft when he said, "We sure are putting out a lot of smoke today." Captain Bob looked back and realized that something was drastically wrong. We were on fire! Black smoke was billowing from the engine room air vents as the

---

2   The U.S. Coast Guard routinely operates throughout The Bahamas, and vessels in distress call upon them, even within Bahamian waters, for emergency services. U.S. Coast Guard vessels are painted with a large red stripe at their bow, making their identity unmistakable.

*Captain Bob called out a "Mayday" on the ship's radio from the
smoke-filled wheelhouse of the Rio Grande.*

boat came to a halt. The engine hatch was popped and immediately
more black smoke burst from the engine room. I was down below in
my cabin when Bobby Heater came running along the gangway to

Sam Staples peers up from the Rio Grande's engine room, during the fire. SCUBA gear permitted the crew to fight the fire with extinguishers.

gain access to the engine room from inside. A soon as he opened the engine room hatch the fresh air fed more oxygen to the fire, which made the flames jump up towards the door. We closed the hatch immediately, to prevent any more airflow to the engine room fire. Up on deck, Sam and Captain Bob had donned SCUBA tanks so they might breathe while in the smoke-filled engine room. Only one person could go down at a time, so they were taking turns fighting the fire. I had grabbed my camera and was clicking off the shots as fast as I could take them. At one point Captain Bob was on the radio shouting, "Mayday, Mayday, Mayday, this is the *Rio Grande* and we are on fire 50 miles off the coast."

In the meantime, the radio officer on board Penney's ship, the *Scandinavian Sun* had heard our Mayday on the radio. He asked her, "Penney, is Randy's ship the *Rio Grande*?" She answered yes and was then told we were on fire. So, Penney was hearing the entire episode unfold over the radio and had tried to break in a few times only to be told, "Not now!"

It seems the fire had started in one of the superchargers on

Ian Kelso is seen here standing by with extinguishers during the fire.

one of the 1271's. Sam and Captain Bob were playing the heroes, asking what the rest of us were doing during the fire. Sam took a moment to give me a hard time about taking pictures. It didn't matter that only one person could get down in the engine room at a time and the fire was soon out. We still had power from one engine as we limped into Palm Beach where the repairs would take place. I wasn't a big fan of Sam, and he wasn't a big fan of mine. I thought he was afraid of the ocean frankly, and treated us like a bunch of boy scouts. He would attempt to give us fatherly advice and we could give two shits. His favorite inspirational story was telling us we were like the fingers on a hand, and that we all had to work together in order to be efficient, "like a hand."

The boat was down for months; the damage was close to a million dollars. All three engines had to be pulled out, gone through, and the engine room cleaned up. It was a big, detailed, time-consuming job.

I got some serious satisfaction when Sam asked me for the pictures of the fire. The authorities and the insurance company were very keen on seeing the pictures to back up our story and make sure it wasn't an insurance job. Now Sam could show them that what he said was true, and he had pictures of himself saving the day!

We docked the boat at Performance Power Systems in West Palm Beach to get the repair work done. The area was shady and close to the railroad tracks where you could get anything you wanted. Late one evening I heard a commotion on the boat. The guys had just returned with a hooker. Sam was back in Melbourne Beach and now the boys were being boys. The door to my cabin flew open and they offered me an opportunity to join in on the evening's activities, which I gracefully declined. They were laughing and shouting encouragement and coaching the participates in proper technique. This was getting old and I could see my time onboard the *Rio Grande* was going to be ending shortly. I didn't care for the operations guy and he didn't care for me. One morning the traffic was horrible, and I came in late from my night with Penney in Miami. It was all the reason Sam needed to let me go, which he did, and it thrilled me. I'll never forget Mark Wolf sitting on the deck looking down at me as I packed up and waved goodbye with four digits instead of five. That had both of us laughing, remembering Sam's stupid-ass stories. I shared my experiences aboard the *Rio Grande* with Rex when I spoke with him. To my knowledge, that was the beginning of the end for the *Rio Grande*. I would see Mark Wolf occasionally, but I never saw the rest of the crew again. I married Penney, my new girlfriend, in Louisiana that September.

Chapter 19
The Island

Being back home and off the *Rio Grande* was a relief. Even better, I received a call from Walter. He was working on an agreement with Eastern Steamship Cruise Line which would soon become the Admiral Cruise Lines. The cruise line had recently leased a small island in the Berry Chain of The Bahamas called Little Stirrup Cay. Little Stirrup was the northernmost cay in the Berry Island Chain. It was owned by a man named Neil Ruzic, who had a background in journalism and had built a career as a publisher of a prominent science magazine. When Ruzic purchased the island in the early '70s, it was undeveloped, uninhabited, and was known mostly for its excellent fishing and lush jungle interior. Ruzic had plans to turn Little Stirrup Cay into what he called the "Island for Science". Ruzic wanted to build facilities and bring researchers to the island to work on alternative methods for harvesting food and pharmaceutical substances from the sea, harnessing solar and wind power, and desalinating ocean water. The island was approximately a mile long and a quarter-mile wide, gently sloping upwards towards its center to a height of around 60 feet. The Bahamian Government had denied Ruzic permits for his projects, and so instead he had leased the island to the cruise line. Next door at Great Stirrup Cay, Norwegian Cruise Lines had been running what they called an "Out Island Adventure" for many years. They would tender passengers from the cruise ships to the island where they could rent snorkel gear, Sunfish sailboats, swim, and enjoy a large buffet-style BBQ. This proved to be lucrative for the cruise line as they kept the revenue in-house by not visiting a port-

The Emerald Seas

of-call where the money went to someone else. The NCL (Norwegian Caribbean Lines) snorkel and sailing program employed around 20 dive instructors on its four different ships, all of which stopped at the island. The sailing and snorkeling generated several million dollars a year in revenue for NCL, which would have been lost, had they stopped at a different port-of-call. Admiral Cruise Lines wanted a piece of this action so they leased Little Stirrup Cay, which was directly across from Great Stirrup Cay. Both islands were within sight of one another.

Admiral Cruise Lines had only one ship the *Emerald Seas*, which would stop at the island on Wednesdays and Sundays. This allowed for the island to be cleaned up and ready between ship visits. Walter wanted to try couples this time. He thought they would be more grounded and that couples, unlike the guys, would concentrate on the job and not the girls. I had to bring Penney up to at least Dive Master's certification, so she could get insurance from PADI[1] for the job. Penney was good in the water and had no problems passing the requirements. The second couple Walter had hired were Pat and Kathy, two Canadians, fresh out of dive instructor school.

The plan was for Penney and me to go to Little Stirrup Cay

---

1   Professional Association of Diving Instructors

Above: Little Stirrup Cay looking eastward toward Big Stirrup Cay. Note the trail running the length of the island.

Below: a view of the eastern tip of Little Stirrup Cay looking northward. The villas, dock, and guest beach are plainly visible.

and get acquainted with the workers and the island layout. Penney's job was to train Pat and Kathy on how to sell the program onboard the ship once the island was ready for passengers. My immediate supervisor, Bill Myers, and I flew to Great Harbour Cay, where we took a boat to the island. I found out that Great Harbour Cay had a reputation; yet another island surrounded by blue waters and sunken planes at the end of the runway. Great Harbour had a great airport, a storm-proof harbor, and a failed resort in ruins. Once we landed at the airport, we met briefly with Bahamian customs and immigration authorities. The cruise line had gone to great lengths to secure work permits for us non-natives, so being on good terms with the local authorities was the smart approach. We drove across the island, past a nightclub, several small stores, and a mix of natives and ex-pats. Large vacation homes lined the road along the coast, with smaller native houses clumped together in the island's interior.

We arrived at the community dock where a small boat was waiting to take us to Little Stirrup Cay. Leaving the dock we wound our way along a channel through the mangroves. The channel was basically a tunnel through the mangroves, with lots of blind corners but in no time at all, we found ourselves on open water. Once the island came into view, I could see four medium size stone villas. These were one-room villas, facing out to what was known as Slaughter Harbor. Great Stirrup Cay was visible on the opposite side of the bay. We tied up on Little Stirrup Cay and walked down the dock past an area reserved for native vendors, up a slight incline to a large, round, outdoor tiki bar. Connected to the bar was a large kitchen, with a BBQ grill, beverage coolers, refrigerators, etc. Connected to this was our "Aquatic Shack", which would serve as our dive shop and the location from where we would dispense the snorkel gear. They also had what they called small outpost bars in other areas of the island. The dock extended far out into the water providing the cruise ship tenders with the depth needed to safely approach.

The villas were made from native rock and stone by Bahamian stone workers who can be quite talented when working with local materials. They had one large room with very high ceilings, sliding glass doors looking out to Slaughter Harbor, and a view of Great Stirrup beyond. An incredible view. I inspected the villa in which Penney and I would live. In one corner was a two-burner Coleman stove hooked to a propane tank, and a refrigerator that had a bad door on the freezer so it never closed. This was the kitchen area. There was indoor plumbing for the toilet, for which, we were thankful. On the table next to one wall was a single-sideband radio, which we would use to contact Miami and the outside world in the evenings when atmospheric conditions were favorable.

We continued with our tour of the island, checking the de-salination plant housed behind the villas, then hiking a small road that ran up the middle of the island from one side to the other. The height at the top and center of the island provided a panoramic view of the west bay side of the island. Here were the ruins of a loyalist settlement from the late 1700s. Those folks who continued to support King George III of Great Britain during the American Revolution

Home away from home... the author's stone villa on Little Stirrup Cay was equipped with a single-sideband radio.

A view of the trail which ran the length of Little Stirrup Cay

were known as loyalists. As the war concluded with Great Britain being defeated by the Americans and the French, the most notable loyalists were no longer welcome in the United States and sought to move elsewhere in the British Empire. The crown rewarded many of these loyalists with land grants in The Bahamas. One such family had evidently landed at Little Stirrup Cay, however, no record of them exists today, except for the ruins of a stone house and the rock fences that are still standing. On the east side of the island, waves crashed against the razor rock cliffs that lined the shoreline. This side was exposed to the northeastern winds and was far from calm much of the time. Close to the rocky eastern side was a small freshwater pond, which the previous inhabitants used to pull water from in the past. Depending on rainfall for replenishment, in the past, this little pond had provided the island with fresh water. The island slowly tapered down to a small point at the northwestern tip, surrounded by thick dense brush.

We were introduced to a few of the Bahamian workers who had worked on the island for the previous owner. One man we met by the name of Tony Robinson was from Harbour Island in Eleuthera, which is a long, thin island in The Bahamas archipelago. It's known for pink sand beaches lining its ocean side. Tony was a jack-of-all-trades... he did it all. He ran the bulldozer on the island; he was the mechanic, he could weld, he was a good man to have around. Tony was the only Bahamian to live on the island, and all the others lived on Great Harbour Cay, a few miles away. Tony was Bahamian, but he was treated like an outsider by the other natives. The man in charge of the workers was named Fussy. I soon learned why he had that name because he could fuss up a storm. He was always pissed off. We met Fussy and his son Tawny. Fussy looked down and was reluctant to make eye contact. It was apparent that he regarded us with suspicion. Everyone had a nickname, it seemed. There was Money, and there was Elvis (he was a singer). And there was a guy named Circle because he was one of very few who had a car, and he drove around, and around Great Harbour Island all the time, so they called him Circle. Another worker they called Shoes. The reason being that as a poor kid in school he always wore shoes that did not match. One kind on one foot, another kind on the other. So the other kids laughed at him and called him Shoes. The name stuck into adulthood. Shoes was scared to death of horses. If he saw a horse-drawn carriage while in Nassau, he would go nowhere near it. He would go around the block or take a different route altogether. I asked Shoes why he was so scared of horses. His answer to me made perfect sense. He said he feared horses because their teeth were bigger than his and they could run faster than him! We met Ben who was a pastor, a large, quiet, thoughtful man. I grew fond of Ben, his quiet manner, his respect for everything around him, and his self-assurance. I knew immediately that I could count on Ben if I were in trouble. The only girl in the mix was Maggie. Maggie was a tough, no bull shit kind of woman, loud, confidant, and opinionated. She was large and full of life. She would sell island crafts and help manage other island vendors. Maggie could cook, conch, fish, stingray, or just about any island dish you can imagine.

Bill and I flew back to Florida. He would handle the logistics of getting equipment to the island. I would packing and preparing for the move to the island. I had one more thing to do before I left for the island. Remember the wreck that my friend OP told me about? I had done nothing with the information I had gathered on the side-wheel steamer, *The City of Vera Cruz*. I had heard that the word was out on the wreck, and the dive club from Kennedy Space Center, the KSC Barracudas were helping themselves to parts and pieces of the wreck. The shrimper who originally found the wreck had artifacts from the wreck on display at the 76 Fish and Dive Center. I felt it was a waste to sit on such a great story and the wealth of information I had uncovered. Since I would not be around to dive on the wreck and profit on my research, I called the *Florida Today* newspaper. I met with a reporter and they loved the story. The headline was on the front page of the September 18, 1983 issue of the Sunday edition in the Brevard section. It read, "Shipwreck; Boatload of treasure awaits plunderers of the Vera Cruz." The paper included many of my quotes in the article. "Diving that wreck could be as lucrative as hell," Lathrop says. "She's just waiting for somebody to go out and pounce on her." Nothing like stirring the pot just before I left town! The story was very well-written by Michael Crook. He had well used my research and had written the story of the ship's sinking, and the struggle of the survivors. It was a solid page and a half of newsprint. The article showed a picture of the hatch cover which was on display at the 76 Fish and Dive Center.

It didn't take long for my soon-to-be-disconnected phone to ring, and I received many calls from many individuals. One, in particular, I recall with some amusement. It was a call from the girlfriend of the shrimp boat captain that found the wreck. She chastised me pretty thoroughly, wanted to know what gave me the right to tell the story, and to publish a picture of the hatch cover which was on display at the dive shop. I explained to her that loose lips sink ships and if her boyfriend wished to keep the information confidential, and not have any pictures taken of the hatch cover, then he should not have put it on public display. It was a good thing I was leaving town because I had all the concerned parties stirred up about the wreck. Soon afterward, Lee Spence filed an admiralty arrest on the wreck and claimed it for himself and his new company with my best wishes. Everyone had the opportunity but Lee recognized the potential value and moved on it first. I finished my packing, turned on the house alarm, and headed for The Bahamas via Miami.

I boarded the *SS Emerald Seas* in Miami, a ship with a rich history. Originally the *SS Emerald Seas* had been the *USS General W. P. Richardson* (AP-118), a troopship that served with the United States Navy in World War II.

It was in October 1972 that she was sold to Eastern Steamship Lines Inc, of Panama, who renamed her *Emerald Seas* and she received some modifications, with her tonnage now officially being registered as 18,936 GRT. On December 8, 1972, she began the popular shorter three and four, or seven-day cruise circuit, sailing out of Miami to The Bahamas. She had a section added to her midsection and afterward had a few degrees list to the port side, so many people referred to her as the ship that tilted.

This ship differed greatly from the Scandinavian Cruise Line ships, as she had a seasoned crew and years in the business. The staterooms were older, therefore much bigger than many of her competitors. My wife Penney was on board with me and she would work with me on Little Stirrup Cay until the island was ready for cruise customers. The main *Emerald Seas* market was the 3 and 4-day venue cruising from Miami to The Bahamas. The 3-day cruise started on Friday afternoon and returned on the following Monday morning. The 4-day cruise embarked on Monday afternoon and returned to Miami the following Friday morning. The weekend cruises were always full of party-goers, while the 4-day cruise clientele was more laid back.

Penney and I settled in on the *Emerald Seas* for a few weeks to become accustomed to the ship and get our shore excursion office ready for business. Our cabin was much improved, it even had a porthole, which is a huge benefit. We had several dining options, from petty officer's mess to the engineer's mess, and passenger's dining if we wished. The cruise director was known as Gino, and he'd been on this ship for close to 10 years. The assistant cruise director was Jody. Penney had worked on ships much of her adult life and she settled right into the routine, making many friends along the way. Everyone loved her southern bell accent and her southern charm.

We would leave Miami on Monday evening, arriving in Nassau on Tuesday morning for the day. I got to know Nassau well, especially a little tavern named the Green Shutters Inn, which was established in 1814 but regretfully it has been closed for many years now. I have so many special memories of this tavern. I would take our new diving instructors to the Green Shutters when it was their first time in Nassau. A couple of pints of British Courage Ale in their belly and they told me everything I needed to know about them, and more. The Nassau library was just up the hill from the tavern, which I would visit frequently, trying to find answers to questions that would arise on the island. One of my favorite places in Nassau was Fort Charlotte, a 100-acre, limestone fort with moats and bridges, built in 1788. To get to the fort, you walk down Bay street past all the souvenirs, jewelry shops, and the straw market. Dodging the weed dealers on the street as I went. A pain in the ass. Most of what they were selling wasn't weed, it could be tea or just leaves from the ground put in a baggie. They would approach me every time asking, "Something for the head?" I would always respond, "No thanks, I got a hat." This would always confuse them momentarily, enough time for me to get on down the street. It was worth the walk. The view on top of the hill looking out from Fort Charlotte down to the blue waters was killer.

This ship differed from the Scandinavian ships we had been on in the past. The *Emerald Seas* had an older clientele. Many repeat customers. Different ships attracted different cruisers, depending on age, itinerary, and what they were looking for on their vacation. The four-day cruises were more laid back, and the three-day cruises were more for the party-goers. The ship had two formal Captain's cocktail parties on every cruise. The captain, officers, and department managers would form a reception line, all in formal attire for the event. Penney was at the front of the line and always looked beautiful, with her slender features, and her long dark curly hair falling to her shoulders. It was her job to introduce each couple to the captain and staff. I think they enjoyed her strong southern accent. I also had to stand in the reception line because I was the aquatics manager. Each couple would pass down the line exchanging formalities with the captain, officers, and staff. You never knew what some folks regarded as formal at these parties. The surprises were many. It might be a bridesmaid dress or something they last wore at their prom. It was difficult to keep a straight face all the time. Both Penney and I came to loath the cocktail parties. The first was at 4 o'clock, and then another at 5 o'clock. They served just three types of cocktails; Martinis, Manhattans, and Whiskey Sours. It could be dangerous. It was tricky to watch your consumption and your mouth during the second party. I came to love martinis and still do. The ship's itinerary after Nassau was Freeport, Grand Bahama, but this was changing with the addition of the island. The ship would no longer stop at Freeport — it would sail to the Little Stirrup Cay and spend the day there. The cruise lines understood that this was an opportunity to keep all passengers on board the ship spending their money with the ship instead of spending it shore-side. They kept more revenue to themselves instead of it going to the shops in Freeport.

After several weeks on the ship, we were transferred to the island via a small launch from the *Emerald Seas* anchored offshore. This unpublished stop brought a small crowd to the railings to watch us depart. Penney and I took up residence in the middle stone villa, Tony Robinson, the island mechanic, lived in another of the villas, and John, an Englishman who would run the tender operation, lived in the third villa. No workers lived on the island aside from Tony. The other workers would arrive every morning in a boat from Great Harbour Cay. The villa amenities were very "basic", as I noted on my first visit. What I didn't see on my first brief daytime visit, only came out at night. Rats. They infested the island. The island had a cat the first week until it went back home to the states with the owner of the island. At night we could hear the rats running along the trim just above our heads. On several occasions I flipped on the light with pole spear in hand, engaging in battle with a few of them, but it was going to take a prolonged war, not just a battle to kill all of them. One trick I discovered by accident that worked well were garbage cans filled with water about one-third of their capacity. The rats would jump in and drown... the stupid ones anyway. The rat's only enemy on the island was a resident screech owl. The owl would fly low at night down a path we often walked and would startle us. Ben, the pastor, held great reverence for this bird. I don't suspect that there were many owls around, so it made this one even more special.

I started working with the crew on several projects, or rather, I tried to work with them anyway. The crew was Money, P John (a Haitian), Tawny (Fussy's son), and Elvis. It didn't take long for me to notice that when I gave instructions to do something, they would look at Fussy before they reacted. Fussy would scowl at me and nod his head in approval to the others. After a few days, Fussy called me over to the side and said we needed to talk. He looked me in the eye with sweat dripping off his forehead, and said, "Me no walk like crab". He went on, "Me no walk like crab, me no walk sideways. Those are my niggers. If you want those niggers to do something, you ask me. You understand?"

I looked at Fussy, and responded, "No problem, Fussy." That moment changed everything moving forward. Fussy reminded me I was an outsider, and that there was a hierarchy in place and a proto-

The author is seen here cleaning Conch (pronounced "konk") with his Bahamian crewmates on Little Stirrup Cay. Bottom left to right: Elvis, Fussy with hammer in hand, the fellow wearing the white T-shirt is Money and the younger fellow standing is Tawney. To the author's immediate left is Mark.

Looking out toward Slaughter Harbor from the guest beach facilities on Little Stirrup Cay

col I must follow. I had no problems from that moment on with getting anything I wished done. I would tell Fussy what I needed and he would yell and those men would move.

As for the man they called P John... P John was an illegal Haitian immigrant, who had somehow made it to Great Harbour Cay. The Bahamian locals teased him and made fun of him. His proper name was John Phillip, but they called him P John, just to mess with him and make him mad. When P John got mad, he would go into a tirade, yelling, screaming, and making a big scene. So they intentionally pissed him off just to watch him go bonkers. The first time I heard him throw a fit, I came running thinking he was going to kill someone. I found him in a tirade, with the others sitting around laughing at him, satisfied they had done their job of winding him up.

I spent several weeks getting our aquatics shack set up. We had close to 500 sets of masks, fins, and snorkels which required marking and sorting. I had five mini fish sailboats and a half dozen windsurfers, which all required rigging. The rest of the time I spent exploring the bottom surrounding the island. The north side of the island was all razor rock, with steep sides that dropped off immediately into 25 feet of water. The bottom was teeming with life, sea-fans, snapper, grouper, and coral; it was pristine. Following it to the east took you to Slaughter Harbor. The bottom here was shallow and sandy with areas of patch reef, and a large rock or two extending up and out of the water. As you came around Slaughter Harbor to the leeward side, the bottom was all sand and grass. For miles to the west, the horizon blended in with the sky. A little further around to the west side I discovered grass ledges, maybe one to two feet high, and these small spaces were loaded with lobsters. I would return to this lobster hole almost weekly to harvest a few at a time for dinner. You never want to clean out a hole of all the lobster; always leave a few in the hole to invite others to join them. This little spot offered another experience, to my surprise.

The ruins of the original Loyalist house on Little Stirrup Cay are seen here behind a wooden barrier.

The first time I pulled a few bugs from these grass ledges, I didn't have a bug bag, so I had a lobster in each hand. It was shallow here, about chest-deep. I had a lobster in each hand and they were just flipping those little tails, snap, snap, snap. When I looked out I could see I had lots of company, Black Tip sharks. A soon as the lobster

flipped its tail the sharks started swarming, one, then two, then a dozen, in minutes. These guys were about two to three feet on average, and fast! They would dart in and out like a pack of dogs. The first time this happened I was awe-struck! From then on I saw them every time I went to grab a few lobsters from that hole. I could time to the minute when the sharks would show. Far into the future, when a new instructor would arrive on the island, I would ask them if they wanted to see some sharks. I would take them to my little lobster hole in chest-deep water and ask them to just stand there and watch. They almost always thought I was joking or messing with them. How dangerous could it be in four feet of water? I would swim over to the ledge, grab a lobster and hold him upside down in the water. The lobster would flip that little tail, snap, snap, snap. Just like ringing a dinner bell, the sharks would come running and my new instructors would shit. I had so much fun showing people my little Lobster hole!

The conch were plentiful on the grass flats and we ate our share. The locals showed me how to knock them out of the shell and clean them. I was soon as good as any of them at knocking them out of that shell. I make great cracked conch. Roll those rascals in corn-flakes, and fry em up, good!

Above: the author on the rocky north shore of Little Stirrup Cay

Below: on occasion, the weather would keep the Emeral Seas from anchoring at Little Stirrup Cay. Here, on one such blustery day the author photographed the fresh water impoundment facility.

The Bahamas is noted for its pristine waters, and the diving is spectacular as seen here on the north shore of Little Stirrup Cay.

We would spend our nights looking at the stars in the unhindered sky. Watching twinkling ships' lights on the horizon and listening to the single sideband radio. The radio was entertaining; the calls from South America in particular. Coded calls asking "if the cattle were ready." We suspected that many of the calls had nothing to do with cattle; we knew what they were moving, and it wasn't cows. Then you would hear the ship-to-shore calls from the cruise ships. People are stupid. We would laugh and marvel that the people on these calls never considered that they had a vast audience listening to them in the dark. "I love you, honey, I love you, darling." "We love you too!" Folks talking about all kinds of personal business they shouldn't be talking about.

One day, Tawny, Fussy's son, came running to me. "Randy, there are some white people on the island." Tawny and I both knew white people meant strangers, outsiders. We allowed the natives to come and go as they wished, so when he said "white people" I knew they weren't locals. I could see a small group of people sitting at one of the many picnic tables we had on a rocky point. It looked to be about six people, two white guys, two Bahamian girls, and two Bahamian dudes. As I got closer, I recognized the two Bahamian dudes, one was customs and the other was immigration. I greeted them with open arms once I recognized them. They introduced me to the first of the two white guys, Klaus. Klaus was German and had a heavy accent. He sounded like a mix of Yosemite Sam and Sylvester the cat. The second white guy was a Colombian named Carlos. He was a pilot who lived in Miami. They had all been drinking, and they were pretty screwed up. They had a cooler full of liquor and beer and were having a good time. I accepted a beer from them and we engaged in casual conversation. Klaus kept looking at me, sizing me up. He soon looked me in the eye and said, "Pedico?" I knew the word... the word means "cocaine". Klaus looked at me again and asked, "Pedico?"

I looked at the customs and immigration dudes thinking, "What the hell is going on here?"

I looked at the customs guy and then back to Klaus. Klaus laughed and looked at me, then looked at the two Bahamian officials and told me not to worry. He said, "No worry, I pay." He then made a gesture of peeling money off a stack and handing it to them. He repeated this several times.

The customs guy looks over at Klaus and says, "Klaus, you got a big mouth." Carlos just sort of looked on, amused by Klaus and his antics. Klaus passed a small bottle with the coke in it to the girls who helped themselves. I kindly refused the offer. Klaus then spoke about the west end of our island, how nothing much went on there, asking me if that wasn't correct. I agreed with him that it was pretty quiet on that end of the island, mostly jungle. I knew where this was going. It became obvious they wanted to use the west end of the island for their business, storing and staging product there until they could pick it up for the next leg of its trip. They wanted to confirm what days the ship would stop at the cay and for how long. Their last request was for us to leave the keys in our Herald Haulers, which were small four-wheel motorized carts we used to haul stuff around the island. I was trying to feel my way around this situation. After much discussion, I explained to them I didn't care what they did, didn't wish to know, didn't wish to take part. But I said that by accident workers often left the keys in the haulers, and to be careful as I never knew who might show up from Miami.

The group soon departed for Great Harbour Cay in the speed boat they had arrived in earlier. Both Klaus and Carlos insisting that we come to Great Harbour Cay to have dinner and party with them. I explained we didn't have a boat, and they said no problem, they would send a boat for us. It wasn't long before they kept their promise and sent a boat for Penney and me to take us to Great Harbour Cay.

Carlos and Klaus had an enormous three-story stone mansion on the island they had leased. Klaus was a character. He wore a bullet on a gold chain around his neck. The story he told me was that one of their planes with a load on board had been intercepted on one of the islands. The authorities had chained the prop to prevent the plane from taking off. Klaus and Carlos had removed the chain from the prop and were just taking off when the police arrived. The police fired their rifles at them when they just got in the air, the bullet went through the plane and hit Klaus, falling to the floor. Klaus screamed ouch and bent down, picking the bullet up off the floor of the plane.

He had it mounted in gold and wore it on a fat gold chain. They wined and dined us that evening. They took us to the island disco, which had a large crowd and was hopping. Penney felt obligated to dance with every Bahamian official who asked her to dance, so she was on her feet most of the night. It didn't take long to figure out everyone was on the smuggler's payroll. Everyone. If they weren't on Klaus's and Carlos's payroll, it was another group's payroll. Little Stirrup Cay had a long history of smugglers and smuggling, going back to US prohibition. In the '70s, the police had cornered a group of smugglers on the island. To flush them out, the local authorities set the woods on fire. The culprits escaped with no trace, but police found two bodies and a Thompson sub-machine gun from the 1920s in the ashes. History repeats itself.

The day arrived and the *Emerald Seas* started anchoring off the cay and disembarking passengers ashore to their very own private out island. The passengers loved the island, the BBQs, the outpost bars, sailing, snorkeling, windsurfing, crab races, you name it. The ship would stop on Wednesdays and Fridays, giving us several days to clean up and prepare for the next visit. I continued to explore the island, especially the ruins on the hill which most likely dated back to the late 1700s. I poked around the stone ruins of the house and followed the rock fence throughout its length. I found pottery, glass, and iron, which was enough to keep me entertained. At the foot of one of these walls, I buried a 25 caliber Sterling semi-auto pistol in case of a dire emergency. Guns are illegal in The Bahamas, but tell this to the bad guys. On the island, I only had myself, no police.

The program was doing well, revenue was doing well, and the cruise line was happy. Ordering food for us on the island was fun because we got the same provisions provided to the ship itself. The only problem was that everything came in large number #10 cans, restaurant size. I would order enough to share with the workers. We would divide the large cans among them. They had never eaten Cornish hens before and they became a big hit with the local crew. Sharing these things with the locals made a big difference in their lives. I quickly learned that if I watched out for them they would watch out for me. Our sister island, Great Stirrup Cay, was leased from the Belcher Oil

company of Miami, by the Norwegian Cruise Line, and there were no full-time occupants on the island. Norwegian Cruise Line ships only stopped there three days a week.

When the protected lagoon was empty, Windjammer Barefoot Cruises would bring in their ship, the *Fantome* to party. They would send a small skiff to fetch us so we might join in the festivities. The *Fantome* had a rich history, originally ordered for the Italian navy, she was purchased before completion by the Duke of Westminster, who finished her as a yacht (launched in 1927). Westminster only used her a few years. In 1969, Windjammer owner and founder, Michael Burke, flew to Greece to purchase the schooner directly from Aristotle Onassis. He bought her, sight unseen, in exchange for a freighter. Windjammer then set about refurbishing *Fantome*, which became the flagship of their fleet of six vessels. She was a beautiful vessel; the crew allowed me to roam her upper and lower decks, taking in the wonderfully done restoration efforts that returned her to her former glory. Windjammer knew how to party, and clothing was often optional. I think of the *Fantome* and her crew often. On October 24, 1998, the *Fantome* departed Honduras for a planned six-day cruise. More than 1,000 miles away, Hurricane Mitch churned and was expected to pose a risk to Jamaica and possibly the Yucatán Peninsula. Captain Guyan March decided to play it safe by heading for the Bay Islands and wait for the storm to pass. The following day, Mitch seemed to change course. *Fantome* immediately changed course also for Belize City, where she disembarked all of her passengers and non-essential crew members. The schooner then departed Belize City, first heading north towards the Gulf Of Mexico in order to outrun the storm. Mitch changed paths several times and each time the *Fantome* adjusted her course accordingly to avoid the storm's fury. As Mitch moved in on Roatan and Honduras, *Fantome* made one last desperate attempt to flee to safety, now heading east towards the Caribbean. Mitch's forward motion picked up, and she could not outrun the storm. Around 4:30 PM on October 27, 1998, with Mitch having weakened but still at Category 5 intensity, *Fantome* reported she was fighting 100-mile-per-hour winds in 40-foot seas. They were just 40 miles south of Mitch's eyewall. Radio contact was lost with the *Fantome* shortly after that. All that was ever found of the *Fantome* were life rafts and vests labeled *S/V Fantome* off the eastern coast of Guanaja. All 31 crew members aboard perished. Those of us who have spent years salvaging the past should never forget the tragedy and heartbreak these events had on individuals and families. 679 tons of wood, steel, and canvas went to the bottom of the sea.

As winter approached, the weather systems would bring strong winds and rain sometimes for days with the seas being so rough that often the *Emerald Seas* could not stop to drop off passengers or resupply us with provisions. We had periods that stretched into a couple of weeks without supplies. I had my pole spear, so this was never a problem. Fresh fish were abundant. The conch were plentiful too. We would construct a "conch corral", a circle of rocks that the conch couldn't crawl over to escape. We would fill it full of conch and when we needed one we would fetch it out of the corral. I learned to identify many of the plants on the island. We had goat peppers we used in conch salad, Bay Rum (pimenta Racemosa) which we used to make tea. One morning I saw the workers eating berries off of a particular tree. I asked them what the berries were and they laughed and told me they were "stop berries" (Pigeon-Plum). They must have had a terrible night because they explained they were used to remedy diarrhea. We now had watermelon growing everywhere on the island after feeding it to the tourist for six months who'd been spitting the seeds everywhere.

The rat problem had grown worse, and the cruise line had ordered that poison be used to rid the island of them. It was a poison dispensed in wax-like cakes, impregnated with corn. The rats would eat it, drink, and die. The rats had only one natural predator on the island, the island's sole screech owl. Once the poison took hold of the rats, I never saw the owl again, I suspect the owl ate one of the poisoned rats and was dead too. This made me extremely sad because I had become accustomed to seeing the owl in the moonlight and hearing it in the darkness. The island was slowly changing, one step at a time, leading to its demise.

We now had sailboats stopping and anchoring in Slaughter Harbor, sometimes overnight, other times for a month or two. They would eventually find their way to our dock to check out what was happening. We would give them a quick look-over and decide whether we wished to be civil to them or tell them to go away. The few times we were friendly, we usually ended up being sorry. So, over time we took a fairly hard stance: no one allowed on the island, with one exception... they could come ashore on passenger days and drink at the bar. Most of the blow boaters were tourists, but others were not. It became apparent that many of them were waiting to rendezvous with someone else. They were waiting to take on a load in the darkness when the time was right, and they would disappear as suddenly as they appeared. One evening a cigarette boat came screaming into the harbor and up to our dock. Onboard were three Spanish guys with bandannas around their heads. They shouted out, "Donde esta Chub Cay (where is Chub Cay)?"

I pointed to the south and replied, "Es por ahi (it's that way)". They waved and sped off towards the south. I laughed and wished them luck.

One day I look out at the harbor and I see this 30-foot motorboat coming into the harbor using the wrong entrance, which was shallow and rocky. I'm thinking, "Who is this idiot?" With some luck, he weaved his way through the cut and up to our dock. I walked to the dock and there standing on the boat is Klaus, all happy and full of himself. He jumps on the dock, hugs me, and asks me how I like the boat. I told him it was a nice boat. He said, great, because it was my boat, his gift to me to use as long as I wanted to use it. He then showed me the boat's little cabin and its communications. He showed me three different lengths of antennas explaining, "This is a two-meter, this is a three-meter," etc. Suddenly I understood where this was going. He wants me to be his eyes and ears via the boat and radio. I'm in a tight spot. I would have loved to have access to this boat, but not with the stipulations attached to the deal. I explain to Klaus that it's an extremely appreciated offer and I'm thankful, but I can't accept his gift. I tell Klaus that if I suddenly had this boat it would really draw the attention of the office back in Miami! I wiggled and squirmed my way out of accepting his boat. I said it was a grand gesture, appreciated, but one which would have everyone suspicious, and just bring him trouble down the road.

We became more and more aware that under the cover of darkness there was lots going on around us. One night the sound of a large plane awakened us as it flew low overhead. It sounded like a large plane — big and low. Penney and I walked to the top of the hill to get a better view of what was happening in the darkness. We watched as maybe 6 or 7 boats got in line behind one another, spaced out several hundred feet from one another, with their lights on for the cargo airplane to see them below. It looked like a runway. Once all of them were in a line, the plane would line up with the lights, and start kicking out bales as they flew low overhead. The front boat would peel off and grab a load, while the plane circled and dropped another load for the boat behind the first one, and so it went. By the time the last boat in line picked up their load, the first boat had unloaded and was back in line again. This went on for over an hour. Once they loaded, they would run like hell in the darkness to Cistern Cay, unload, and return to the line-up for another. This system eventually proved to be deadly. In another year, two young Bahamians who were the sons of a prominent local politician crashed into each other at high speed in the darkness, killing both while picking up bales at night. These deaths spurred a Royal Commission of Inquiry by the authorities that ended the large scale of the smuggling in that area, but new smugglers soon took their place.

When we had our work done getting the island ready for the next passenger day, we would snorkel all around the island. Penney had become very comfortable and skilled in the water in a short time. One calm day we grabbed our gear and swam right off our villa past the three large rocks in our snorkel area out to deeper water. There was no wind, not a ripple on the water, the surface flat, the water clear. We had swum out to where the bottom was around 60 feet below us. The visibility was almost unlimited, I could easily see a hundred feet or more. Penney had her fish identification card and was floating on the surface, looking at her card to identify a fish that had caught her eye. I was practicing my breath-holding and free-diving to the bottom followed by slow ascents. It was an absolutely beautiful day, the sun's rays almost reaching all the way to the bottom. We each kept doing our own thing, enjoying the day, not paying much attention to anything in particular. I dove to the bottom and was slowly rising to the surface when I looked over at Penney and my heart jumped in my throat. Penney was still floating motionless on the surface, looking at her fish card. About 20 feet below her was the largest bull shark I have ever seen, then or since! It was also motionless and had turned on its side, looking up at Penney. This shark was an easy 15-footer, and it seemed almost as big around; it was huge! I surfaced trying not to panic and told Penney to look under her. When she looked down and saw the shark directly under her, I heard her starting to cry. I slowly swam over to her, told her to get behind me and swim for the closest rock. Penney did as she was told, facing the rock behind me, while I

The author's wife, Penney, lounging on the guest beach between visits of the Emerald Seas. She and her husband had lots of time to explore the surrounding waters when not entertaining the ship's passengers.

Left to right: the author, Pastor Ben, and Brad Young cleaning a load of Conch.

faced the bull shark which had now straightened out and was slowly following close behind us. We swam several hundred feet for what seemed like an eternity. I never took my eyes off the shark, and the shark never took his eyes off of us as we swam slowly for the nearest rock protruding above the surface. Trying not to swim like hell as fast as we could to safety was excruciating. If we had panicked, we would have sent the exact wrong vibe to the shark and might have triggered an attack. The enormous shark was slowly gaining ground on us, suddenly our flippers felt the sides of the rock below us and we scrambled on top of it as fast as we could. The shark had followed us all the way to the rock and was only 25 feet behind us when we crawled up to safety. To say it freaked Penney out is an understatement. I had spent years in the water and my heart was still racing in my chest. Penney had only been doing this less than a year, and it would have a lasting impact on her. We sat on that rock for a long time, waiting for the fear to subside and the shark to go away. I spent my time trying to calm Penney and assure her we would be fine. Just another adventure. It freaked her out, and she wanted nothing but dry land under her feet. We sat on that rock about 200 yards away from the safety of the beach for at least an hour. We stared out on the waters surrounding us, looking for any signs of the enormous shark. I convinced Penney that the shark had gone.

We entered the water once again, and this time swam like hell straight for the beach. It was late afternoon as walked up onto the shore and back to the villa. It had been quite the day! My wife was almost eaten by a bull shark. From that day forward, Penney didn't like her job anymore. When we had snorkelers in the water, she would now only swim out to a large rock, sit on it all day, all the time looking for that shark to come back. The tourist in the water would ask her what she was looking for. It wasn't long afterward that she refused to get in the water, period. I did everything I could to keep her busy doing other things, but she had lost her heart for diving and the island.

Christmas arrived on the island along with a chilly wind blowing out of the north. On windy days I loved to take one of the Mini Fish sailboats out on the leeward side of the island and sail in the strong winds. I would stay just within the lee of the island and tack back and forth. It was cold this day, and I put on a full wetsuit and went sailing in the steady north winds. I did well for about an hour but ventured outside the lee of the island and could not tack back against the now stronger winds. I was being blown farther offshore onto the flats west of the island. I had no choice... I had to try to swim back to the island. When I jumped off the little sailboat into the water, I floated like a cork because of the full wetsuit. I was in trouble, big trouble. I was on a course that would take me miles and miles out on the barren sand flats that lie to the west and went for miles to nowhere. The seawater mixed with the winds and I was gulping for air but getting more water in the wind than air. I was moving fast across the top of the water. I soon saw a sailboat anchored close by. As I came flying by the anchor line, I reached out and grabbed it. I was able to get myself to the side of the boat, where much to my luck hung a line off the side with a bumper attached. I grabbed the line and pulled myself aboard the boat. I lay on the deck and vomited saltwater and gasped for air. I came very close to drowning that day. Fortunately for me, one of

The author is seen here with one of the Mini Fish sailboats at Little Stirrup Cay.

the workers saw me go floating away and had gone and got Ben the pastor. I lay on the deck and puked sea water until I had no more inside me. I heard a small boat coming in my direction... it was Ben. I was so relieved to see that Ben had come looking for me. There was no one on the sailboat I had climbed aboard, and as I got in the boat with Ben, I felt bad for the mess I had left on their deck.

It scared Penney when she heard I was in trouble and she was waiting for me on the dock. I felt really stupid and was ashamed I hadn't been more cautious. I almost lost my life that day. Later that evening I was sitting with Elvis and Money enjoying some liquid Christmas cheer when Elvis suggested we sing Christmas carols. I thought Elvis was kidding, so I said, "Sure Elvis. You start it off." Elvis became very solemn and started singing in a low reverent beautiful voice, "Silent night, Holy night, all is calm, all is bright." I looked at Elvis sitting there in the dark singing "Silent Night" and it almost brought me to tears. "Silent Night" has never been sung before or since with such grace and reverence. I never would have imagined that such a voice filled with so much love and respect would live in Elvis. That Christmas I learned two valuable lessons. Don't be stupid around the ocean, she can take you in a heartbeat. The second lesson I learned was never to judge a book by its cover. Elvis, wherever you are, I hope you're still singing on Christmas.

It had been just over a year since we first stepped on the little cay. Penney and I had earned a vacation and were given a couple of weeks off. We fled back to Florida to watch TV and eat ice cream late into the first few nights. It was good to get back to civilization. Too soon we found ourselves returning to Little Stirrup Cay to relieve Pat and Kathi, the Canadian couple who stood in for us while we were on vacation. I felt a change as soon as I stepped on the dock at the island. Pat and Kathi returned to the ship to run the program on board when Penney and I returned to the island. Everyone seemed pissed off when I returned so I started asking questions. The workers told me what had been happening on the island in my absence. The story was that as soon as Pat and Kathi arrived on the island they took control by placing padlocks and chains on all the refrigerators, freezers, buildings, and anything else they could lock up. The workers took this as an insult to their integrity; they were highly offended. The natives were proud people and had every right to be hurt by these outsider actions. I gathered everyone around and we walked the island collecting every padlock and every chain. We then all gathered at the end of the dock where I threw all the locks and chains into the ocean to the cheers of all the workers. They came and patted me on the back, and thanked me. The Haitian P John gave me the best compliment he could that day. Haitians are superstitious and avoid certain things, like funerals. P John told me that day that he liked me so much that when I die he would come to my funeral. Now for someone who believed in zombies and black magic, this was quite the testament!

I could tell that Penney wasn't that happy to be back on the island. I really couldn't blame her. For Penney, the magic swam away with that damn Bull shark that day. It wasn't too long before she took a boat trip to Great Harbour Cay and made a phone call to Miami to tell them she was resigning from her position. I went back on board the ship and sailed to Miami to see Penney disembark for the last time as a cruise line employee. I needed a break also, so I sold the program onboard the ship for several weeks for a welcomed change, especially the food!

I received news from Walter that Aquatic Adventures had secured another contract with another cruise line and another island. This cruise line was Premiere Cruise Lines, which would sail out of

Port Canaveral, and their island was a small cay very close to Nassau. They expected me to get this program up and running soon with the promise that I would be shore-based at Port Canaveral. This would mean I could spend time home and would only be required to sail a few times a month. I would have an office shore-side. I might even save my marriage, which was slowly fading away.

I went back to the island after a welcome respite on board the *Emerald Seas*. When the ship called on the island, we'd have as many as 300-400 snorkelers in the water. We had an outstanding safety record, and I was thankful and proud for no accidents or drownings. That almost changed one day. We had what we called an "advanced snorkel tour", which would take us on a tour in deeper water along the steep rocky shoreline on the northeast side of the island. I kept the tours to only 10 people in a group. I usually gave a small talk about what to expect on the tour and some do's and don'ts. During my talk, I had one fellow who just didn't want to hear anything I had to say and was such an annoyance I wrapped up my talk early. We equipped all of our snorkelers with a mask, fins, snorkel, and safety vest that inflated manually or by a CO_2 cartridge. If you pulled a cord, it would inflate. Once I had the people in the water I would line them up and they would follow the leader. I would swim at the rear of the group occasionally to make sure everyone was OK, and verify my headcount. We were well into the tour along a very rocky part of the coastline, and this area was full of marine life. I turned and noticed a middle-aged man submerge under the water. It just didn't look right. I swam to him and he continued to slowly sink towards the bottom. I grabbed him by the vest and pulled him to the surface, which wasn't easy as he was a large man. I pulled his CO_2 cartridge cord on his vest to float him but nothing happened: it had already been pulled. I started to blow into the inflator hose on his vest as he thrashed around on the surface in a panic. Once I got enough air in his vest to float him, I moved behind him and laid him on his back in front of me. I started talking to him in a calm low voice telling him it was going to be OK and not to worry, just relax. I kept talking to him, assuring him we in good shape and would be back on dry land in no time. I got the remainder of the group in front of me and I followed, swimming both of us back to the starting point. I whispered to him and talked to him like a baby until he calmed way down.

The man was grateful to be alive. He thanked me continuously and cried a lot. It rattled me a little but gave me confidence that I was up to the job they hired me to do. A couple of weeks later, my supervisor, Bill, came to the island on business and told me the cruise line had received a letter from the guy I saved. I asked Bill, "Was it a good letter?"

Bill replied, "Not really, he's grateful you saved him, but..."

"But what?"

"Here you read it," he said and handed me a copy of the letter.

The letter was addressed to the Captain. The letter writer stated he was writing for two reasons. The first to say what a wonderful cruise they had, how nice the staff was, and the food was great, and how he wouldn't change a thing! Then he explained his second reason for writing. That he was involved in an incident that nearly cost him his life. He wrote: "Randy Lathrop was close by when I surfaced for what I'm sure was my last weak attempt, and because he was there, I am here. I will be forever in his debt for saving my life and it's not easy to complain about someone who has saved your life, but I feel compelled to tell you and Wills Travel Agency how dangerous this snorkeling trip is". He then complained about every detail of the tour. That he was not told how dangerous it was, he didn't know the tour required a long swim, which he felt was too long and therefore was dangerous. The CO_2 cartridge in his vest didn't fire, and they should have been checked before the tour. Now, remember the guy that was disruptive and so annoying it forced me to wrap up my pre-dive talk? You guessed it, it was this guy. He was such a self-confident smart-ass standing on the beach being a big man that he didn't need to listen to what I was saying about safety and checking your CO_2 cartridges before you got in the water. After several of us drug him back up on the beach, he lay there weakened, scared and embarrassed. Evidently so embarrassed that he wanted the concession shut down. Before I wrote this part, I read the letter again, so I could quote from it. Jerry Cornell's business card was still attached to the letter, so I looked him up online. I found out he passed away suddenly in 2000. I don't know how he died, but I hope it wasn't because he was talking and wasn't listening when he should have been!

Almost 40 years later, I still carry memories of the island with me. I remember how it sounded when the wind blew on windy winter days. I think about how quiet it could be on the cay, just the sounds of the ocean and the wind. The solitude of the place sticks with me. No phone, no TV, no stereo. We had AM radio sometimes; I remember how funny the Bahamian bread flour commercials could be on the radio. I wonder who might still be alive and who has left us. I'd rather not know who is here and who has gone. They all stay the same in my head, they will never age. I would make it back to the island frequently for different reasons, always business. I shot an underwater video for a commercial on the island. I arranged for a wrecked airplane we named the "Smugglers Blues" to be dropped in the snorkel area. The cruise line had several props made to place on the bottom as attractions for the snorkelers. They cast cannons and had a fake shipwreck on the bottom. When the sand began to cover over the attractions, I fabricated a little blower for one of their skiffs to blow the sand off the fake wreck. I sent a man over to install it on their skiff.

I overheard a conversation between passengers after they returned to the ship from the island one day about a large area they saw on the west end of the island where the brush had been cut away. Lying around were soda cans, food wrappers, etc. It was apparent that they cut the brush to cover a load, and that people had been watching for several days. So it looked like the west end was still being used occasionally by the smugglers. The west end of the island spooked me. There was something different about it. I stopped snorkeling that area a long time ago because every time I got in the water there I could hear a very high pitch sound, a constant sound. It wasn't a natural sound, it was a high pitch hum. When I say every time, I mean every time. It was enough that I stayed away from that end of the island. But it didn't keep everyone away, obviously.

Chapter 20
Jumping Ship

The contract was signed between Aquatic Adventures and Premier Cruise Line for Aquatic Adventures to run the snorkeling concession for Premiere. I was being transferred back home to sign aboard the *Big Red Boat*, which was Premier's main ship. She was also called the *SS Oceanic*, built in 1965. She was 38,772 gross tons and could accommodate over 1,200 passengers, with a length of 782 feet and a beam of 97 feet. In 1985, Premier partnered with Walt Disney World, providing seven-night land and sea vacations on the *Big Red Boat*. Minnie Mouse christened the *SS Oceanic* and the event made the cover of the New York Times. Premier was licensed to promote Disney characters on its ships until the relationship ended in 1993. Disney then started its own cruise line in 1995.

I was happy to be home, but I dreaded working on a new ship that was sailing on its maiden voyage. Inaugural cruises were a pain in the ass until they worked all the kinks out. Maybe weeks, maybe months? I wasn't looking forward to it, especially when I opened the door to my new cabin. My new cabin was an inside cabin, no porthole like I had on the *Emerald Seas* and the cabin was several decks down... I almost had to decompress when I came up on deck, it was that far down. It was a third of the size of my cabin on the *Emerald Seas*. The toilet was in a corner and you had to place your legs outside the bathroom to fit on the seat. It was awful. New crew, new cabin, fresh problems. I had a new diver named Peter Harvey; he was from California. Pete did an excellent impersonation of Thurston Howell III from Gilligan's Island. He would shout,

"Quick, first aid, a BMW and a Perrier on the double". I still chuckle when I remember Pete doing Thurston.

My first sail on the *Big Red Boat* was a dinner cruise for the media just offshore of the port. Confusion was everywhere, like I said, new crew, new ship, fresh problems. I attempted to mingle as much as I could, but my heart just wasn't in it anymore. I hoped that my attitude would change, or I'd never make it on this ship. When the pilot boat came to pick up the pilot from the ship after he maneuvered her out of the port, I jumped on board the pilot boat for a trip back to port, which was highly unusual but they allowed me to ride with them this time. Getting off the big ship onto that pilot boat was a trip; the tiny pilot boat jumping up and down next to the behemoth of a cruise ship — dangerous. The following week we set sail for Nassau. Our snorkeling program was known as "Splash Down".

The ship anchored at Nassau for a day's adventure and then moved on, according to the travel brochure, to the powdery-white sands of the "Uninhabited Salt Cay, which you may just recognize as the setting for Gilligan's Island and more recently, the Touchstone movie Splash. Supposedly, Salt Cay once harbored a band of pirates!" Blah, blah, blah. Passengers were given big beach towels and while they indulged in snorkeling, shell hunting, beach volleyball games, swimming, relaxing in hammocks strung between palm trees, enjoying a calypso band, and more on the five beaches that ringed the island. There was a traditional Bahamas barbecue with char-grilled chicken, burgers, tropical fruit, and more. The island was renamed eventually "The Blue Lagoon". This was a temporary location, I hoped because it was almost impossible to get 200 to 300 people in and out of the water because of the small beach and rocky shoreline. There was very little to see on the bottom compared to Little Stirrup Cay.

The success of the *Big Red Boat* did indeed lead the major cruise lines to position ships at Port Canaveral. In the first year of its partnership with Disney, over 60,000 passengers embarked on the cruises, and the partnership was also having a significant positive impact on Premier's non-Disney cruises as well. When the contract with Disney was not renewed, Premier then licensed Warner

Brothers cartoon characters, like Bugs Bunny and Pepe le Pew, so that its ships would keep their family-friendly image, while they simultaneously continued to offer Walt Disney World bookings and other cruise packages. The ship's advertising declared: "The *Big Red Boat*. She'll bring you closer and closer... to each other. The Official Cruise Line of Walt Disney World is America's premier family vacation experience. It's The Bahamas, the Sea, and Walt Disney World, in seven wonderful days that last a lifetime. If you miss being a family, don't miss The *Big Red Boat*." They fashioned the cruise for family appeal, with the focus on kids. When we got to Salt Cay, one of the kids' activities was to find the missing cruise director who's been kidnapped by "angry" pirates.

I'd better look for another job. The writing was on the wall — I knew I wouldn't last long. This was way too "Mickey Mouse" for me.

We didn't depart Nassau until early in the morning, so I paid my last visits to Paradise Island Casino and Cable Beach Casino. I enjoyed sitting at a bar in the Paradise casino and people watch, trying to guess who might be who. Which girls were "working girls" was usually easy, but always fun. The grounds at Paradise Island were lush, landscaped, and lighted at night, always beautiful in the evening. I gambled little at Paradise Island, but I like to play at what-was-then the new Cable Beach Casino.

I headed on over to Cable Beach Casino, wondering what my future held for me. I had a few ideas, but nothing concrete. I obviously wasn't concentrating on my gambling as I was down to my last $20 bill and cab fare back to the ship. As I walked to the exit of the casino, off to my side was the Wheel of Fortune, also known as The Big Six Wheel. I plopped down my last $20 bill on the 25 to 1. The attendant spun the wheel. It spun around, slowing down, click, click, click. I walked away looking over my shoulder, click, click, click. It stopped. Both the wheel and I stopped at the same time. It had stopped on my 25 to 1! I Ran back to the table to collect the $500 bucks I didn't have a few minutes ago! Life was good again!

When I returned to Port Canaveral, I called Walter and gave him my notice. I visited the island a few more times to shoot video and do some contract work directly for the cruise line. I have not seen or heard from Walter since. It wasn't long afterward that Aquatic Adventures lost both contracts. The *Emerald Seas* took the operation in-house and hired Bill Myers, my old supervisor from Aquatic Adventures. Premier did the same eventually with the Splash Down program. Peter Harvey went to work for Premier on board the ship in the purser's office. Bill Myers and I have always kept in touch through the years. He stayed with the cruise line in various capacities for 20 years; it was hard for the cruise line to run him off. He would call me now and then for advice. When the program was taken in-house by the cruise line, they wanted a brand name for the program. NCL Cruise lines called theirs "Dive In". Premier's was "Splash Down". After much thought, I came up with "Sea Trek" and proposed that to Bill. Initially, the lawyers for the cruise line were concerned it was too similar to Star Trek but soon signed off on the new name. Several months later I was sitting in the Miami airport waiting on a flight to Belize when I saw a group coming back from a cruise wearing T-shirts bearing a well-designed graphic with the "Sea Trek" logo. That was a pleasant surprise that made me feel good.

Bill Myer on the left with the author

Chapter 21
Fisheye Productions

The DEMA (Diving Equipment Manufacturer's Association) show was in Orlando and I was off the ship just in time to attend and see all the new technology. The show was at the Orange County Convention Center this time. It rotated between coasts from year to year. I saw several friends and former associates at the show. There was a booth that had a large acrylic tank with bikini models submerged in the tank shooting videos of the crowd outside from inside the tank. This company produced custom-made underwater video housings, and they called themselves "HyperTech". This was still in the early '80s, and the technology was evolving, much of it becoming digital and much more compact. I stopped and watched the girls with the cameras. A sales associate approached me, and he showed me the various underwater housings. I was fascinated with all the new advances and technology. He told me they were quartered in Pompano Beach and made the housings at that location. The company was comprised of three to four young guys who were ambitious, energetic, and cutting-edge technology enthusiasts.

I drove home thinking about those housings. No one had anything like them. They were a game-changer. I could envision many applications for affordable underwater video. When I was in the college photography program, I was also a student assistant at Brevard Community College in the Media Department and was familiar with shooting video. The college gave me the job of videotaping guest speakers. One of the primary reasons they chose me was because the speakers were on a stage a long way from the camera, which required a long zoom lens so any slight movement would be exaggerated. I was the only one who didn't shake, so I got the gig most of the time. I had gone to Brevard Community College because they had an excellent industrial photography program, which culminated in an A. S. degree. I have always shot stills but shooting video was bulky and costly. Also, the quality lagged behind on the video when compared to analog film.

The author attended Brevard Community College where there was an industrial photography program. He's seen here in the school's studio.

Meanwhile, technology was fast developing, and video quality and its cost had much improved. The more I thought about it, the more projects came to my mind. I had spent most of the last two years on an island with no place to spend my money, so I had a nice little grubstake. I did my research, and the best technology at the time came in two parts; the recording deck, and the camera. The camcorders were just making an appearance. I bought a new JVC deck and camera and took it down to HyperTech to have a housing made for it.

HyperTech's shop in Pompano Beach was a busy little place. The orders were rolling in with the advancement in technology. I knew of no one else that had modern video capabilities in an underwater housing except for the big studios. I had written a business plan, done some soliciting by mail and phone, and I had my business cards and flyers printed. I named the new company, "Fisheye Productions", and had another sister company named "GetWet" video. I had an idea I called "Contractor Verification", which was focused on just about any job performed underwater. Having worked several FPL and Southern Bell jobs with Aqua-Tech, burying cable, I knew the end product forced the client to take the contractor's word that the contractor had completed the job to specifications because everything was underwater and the client couldn't see it... unless they hired me. The camera would be their eyes. The camera could be a guarantee and confirmation of a job done correctly. My underwater video service could benefit both the contractor and the client.

The housing and the large light system for the video were manufactured, and I drove to Pompano Beach to pick it all up. I had my first job before they even finished the housing, and I was eager to get started. The housing with the gel-cell battery light system combined was big and impressive. I sometimes referred to it as "R2D2". I used a friend's pool and got everything balanced, as close to neutral buoyancy as I could get it. I was ready for my first job and what a job it was!

Aqua-Tech, the submarine cable trenching business I had worked for many times in the past, had contacted me about helping them out on a patent infringement case. My former boss, George, had unfortunately passed on and a fresh man was in charge. His name was Richard, and he was an Englishman in his late 50s, very proper, with a desire to improve Aqua-Tech and settle a few old scores along the way. Richard had good reason to believe that a former employee had copied and infringed on the company's patent for a cable plow. He had tried, but couldn't get close enough to the plow to confirm his suspicions. To further hamper his efforts, it seemed they always covered the plow up when it was out of the water. The suspect patent infringers were currently doing a job near Pensacola, so Richard sent his employee, Roddy, my old friend, and myself to investigate and get the goods. My friend and brother-in-law, Willard, had even offered to put us up at his place in Pensacola.

We arrived in Pensacola and scoped out the job site where the plow was being used. Roddy knew all the guys who were working the job, as he had worked with most of them at one time. So we had to be careful they didn't see Roddy. We decided we would try it after dark. The plan was that we would take the small skiff the company had given us, out under the cover of darkness, and get as close as possible to the plow. They had left the plow in place on the bottom for the night until they could restart the following day. I felt that my powerful underwater lights would illuminate the plow sufficiently to get the images we needed to prove patent infringement.

It was close to ten o'clock at night as we slipped the boat into the water and headed to the lighted barge with the plow just a few hundred feet away on the bottom. Roddy slowed the boat down as we neared and killed the engine. We floated to a stop in the calm water. In the quiet and darkness, I donned my SCUBA gear, and Roddy handed me my video camera and lights over the side of the boat, trying not to make a sound. I slowly submerged underwater to the bottom below the boat and turned on the camera lights. The lights flooded the entire bottom with super bright white light. I was marveling at just how bright they were when I hear a pounding on the bottom of the skiff from above. I turned the lights off and surfaced just in time to see a light beam from the barge flash on the skiff and Roddy as I hung on to the opposite side. Roddy didn't hesitate and started going through the motions of casting and reeling as if he was fishing. He would fake a cast, then reel, and do it again, and again. Finally, the light went off. The folks on the barge were satisfied that Roddy was just a guy fishing. When the light from the barge went off, Roddy was able to tell me that my lights were so powerful that I looked like a big orange UFO under the water moving around on the bottom. He had pounded on the bottom of the boat to warn me before the barge's searchlight turned on him. I crawled back aboard the skiff, and we headed to the boat ramp.

Willard said he had a bunch of good places for us to eat, but they all must have changed hands before we got there. Roddy and I referred to

Roddy Steen relaxing on the deck of an Aqua-Tech boat

Willard's nightly restaurant quest as the "ptomaine tour". Willard tried really hard but had terrible luck. One evening we were driving down the road looking for a place to eat. Willard was driving and asked us if we might be in the mood for something different. Roddy immediately and without hesitation quips, "No thanks, I'm strictly heterosexual."

Well, this really pissed Willard off, and even more so when I couldn't stop laughing.

Roddy yells, "Hey there's a place, pull over." Wouldn't you know it, it was the best food we had the entire time we were in Pensacola.

We did more surveillance the following day. It looked like the job was ending and they would pull the plow out of the water soon. We would have to get some pictures when they pulled the plow out of the water to place on a flatbed trailer. I saw an immense oak tree not too far away I thought it might be a good vantage point for me to shoot pictures. The next day Roddy and I returned to where the plow is being hoisted out of the water and placed on a trailer. We sneak over to the tree, me with my still camera and telephoto lens. We both climb the tree and take positions on a large limb. We are way up in this tree and staying hidden in the leaves was proving to be difficult. I'm getting my exposure readings and dialing in my settings when I turn to look at Roddy next to me. Roddy had taken two small tree branches and had stuck them between his glasses and his head as camouflage. I looked at him, he grinned at me, and I almost fell out of the damn tree! I couldn't help but laugh and couldn't stop. The workers heard us and soon were standing beneath the tree, looking up at us with curiosity.

"Roddy is that you?" one of them asked.

They busted us. We climbed down from the tree with sheepish looks on our faces as the guys greet Roddy with enthusiasm.

"Hey, Roddy, what are doing up there?"

It was apparent they were happy to see Roddy. We had to come clean, so Roddy confessed to them the nature of our mission. They were a little amused and confused at the same time.

"Hell Roddy, take all the pictures you want, we don't care," the workers responded. We walked over to the plow, and I took all the photos I wanted while Roddy chewed the fat with his old buddies. We got the pictures that Aqua Tech needed, not saying a word about how we got them.

I was able to secure enough work to keep me busy. One job would lead to another, and that would lead to another. I took video editing classes in Orlando at a small commercial studio when I wasn't in the field working. I worked with various marine contractors, insurance companies, and other private companies. An insurance company contacted me about a shrimp boat that had sunk a few days before off Port St. Joe, Florida. When I arrived, I met two guys at the boat ramp who were to take me out to the wreck. They had been contracted to transport me to the wreck only, but they also wanted to dive the wreck.

The day was hot, and the seas were flat calm when we arrived at the large fuel slick on the surface indicating the wreck below. The fuel covered the surface of the water all around us. I was getting my

The author is seen here preparing one of his underwater video systems for deployment.

video gear set up while the other two were donning their SCUBA gear in preparation for the dive. I turned just in time to see one of them jump right in the middle of the fuel slick. He submerged momentarily, then resurfaced, asking for a piece of gear he had forgotten. It wasn't long before diver number one became covered in diesel fuel. Diver number two soon noticed that diver number one was now gagging, covered in diesel fuel, and decided he didn't want to dive that bad after all. Diver number one heads for the bottom while I prepare to enter the water. I sat on the stern, using one of my flippers to clear the diesel fuel away, the slick goes away just long enough for me to make a clean entry into the water. I submerged and the fuel slick closed over me as I descended. The gulf water was crystal clear, I could see the shrimp boat sitting straight up on the bottom like a ghost ship. The insurance company wanted to know if the boat's paravanes had punctured its own hull. A paravane looks like a small steel jet airplane... sometimes they are called "water kites". When deployed on an outrigger, they help stabilize the boat from rolling from side to side. The handling of paravanes can be quite dangerous, exposing the vessel and crew alike to the risk of physical damage. I swam over to the wreck and watched as little globs of diesel fuel slowly floated to the surface from the boat. The outriggers had been deployed, and the paravanes were attached to the outriggers. I swam inside the wheelhouse. The radio mic hung from the radio overhead. There was a *Playboy* magazine and several coffee cups sitting perfectly undisturbed on the small table in the wheelhouse. Nothing looked out of the ordinary. It was as if everything was frozen in time. A pack of cigarettes lay next to the LORAN on a shelf, a cigarette lighter right beside them. This was the strangest shipwreck I had ever been on, maybe because it was only days-old and not years-old.

Writing about this now makes me want to see if I can pull some still shots from the video to show how surreal those scenes were. I shot lots of video before I looked up and headed to the surface. When I was perhaps ten feet beneath the fuel slick, I took my regulator out of my mouth and purged it, allowing a large plume of bubbles to part the fuel slick above me. I then surfaced and jumped up on the swim platform without coating myself or my gear with diesel fuel. I joined diver number one in the boat, who had exited the water in the same poor form he'd used to enter the water originally. He had failed to part the oil slick above him when he surfaced and once again was covered in fuel. He gagged, vomited, and cussed all the way back to shore.

I'm not sure what the sinking circumstance was finally determined to be. I know the boat was re-floated soon after I video-taped it sitting on the seafloor. It looked like an easy salvage job if they got to it in a timely manner, which they did.

During this time Lee Spence published a magazine called, "Shipwrecks". He also wrote articles and advice columns for many magazines. I had a lady who read about me in one of the articles Lee had written somewhere? She was in Tampa and wished to meet with me to discuss a treasure venture. I drove to Tampa and met with her and a gentlemen friend of hers. This lady claimed to be a psychic, and she had a treasure vision. If only I had known before I put gas in the car! So she told me her vision. The wreck was a pirate wreck. This was red flag number two, red flag one was her being psychic. So this wreck was in Tampa Bay, close to the channel near the Sunshine Skyway Bridge.

She said, "We needed to hurry."

"Why do we need to hurry?"

She told me that an extensive flood was coming, and soon the entire Tampa Bay area would be underwater. I thought this rather amusing and stated if that was the case, let's just get to the rooftops of banks and jewelry stores and place some large buoys and lines on the roofs. That way when the flood comes we just find the buoys and dive down for the money and the jewels. She didn't find this very amusing. I wished her the best of luck and we parted ways.

I shared the story with Lee. We were both were so amused about it that Lee wrote about it in his magazine. Well, she read the story and was mad because we poked fun at her. I received another phone call from her, and she was pissed. She threatened me, and she was going to sue me. I asked her, "So you are a psychic?"

"That's correct," she countered back.

"Ok, then, tell me one thing."

"What's that," she asked.

"Who wins the case, me or you," I asked. She hung up. The flood never came, and I wasn't sued.

The television show *Inside Edition* came to Fort Pierce to shoot a story regarding the gold coin find that Danny Porter, and Clay Corderay made near the Cabin Wreck. The reporter from *Inside Edition* was Craig Rivera, an American television journalist, producer, and correspondent for Fox News Channel. Geraldo Rivera is his brother. They wanted some underwater footage, and I was hired to do the job. They also hired me to shoot an underwater wedding in the Keys. I'll get to that later.

It was a sunny summer day near the Cabin Wreck when the filming took place, and lots of folks showed up. Mel Fisher was there, John Brandon, Danny Porter, Clay Corderay, Debbie Brandon, Les Savege, all of us on the beach and in the water getting in on the action. It was decided we would reenact some underwater footage, the only problem being that the only coin we had on hand was a 1622

This hand-full of 8 escudos was recovered by Dan Porter, Clay Corderay, and Les Savege working from a converted Hobie-Cat that was mounted with a small gold dredge at the Cabin Wreck. The author, seen here with those coins in hand, was hired by Inside Edition to shoot underwater footage of a staged re-enactment of the recovery.

Atocha 8 reale. We decided that would work. Treasure is treasure. Clay and Danny, being the stars in this production, swam out from the beach and I followed with the camera. We swam around while Danny and Clay took turns posing with the coin underwater. We finished and swam back to the beach where someone asks, "Who has the coin?"

Danny looked at Clay, Clay looked at Danny... the coin was lost. Now, this presents a potential unique scenario; picture this one! If this coin is found by another salvor on the site, it will certainly confuse them to no end! How did a 1622 coin, the same type as found on the Atocha, end up on a 1715 wreck site? If this ever happens, please refer them to this book.

It was a fun day, all of us goofing off and spending time with Mel Fisher wading and snorkeling around in the surf. I made some good contacts that day with other topside production companies who needed underwater footage now and then.

Not long afterward, *Inside Edition* sent me to the Florida Keys to film an underwater wedding. The wedding took place near Islamorada on a small reef just offshore. The bride and groom had special wetsuits for the occasion and they had rented a Tekna underwater scooter which would serve as their honeymoon escape vehicle. They tied streamers to it to take the place of tin cans tied to a car. Many folks in the wedding party were divers, while others were not.

I hadn't checked the marine forecast because I was in need of the money, regardless of any circumstantial setbacks. The topside production company was shooting their footage first. As the rest of us gathered outside the dive shop, the palms swayed a little and I could tell the winds were picking up. Finally, everyone piled aboard the dive boat and we headed for the reef. As we cruised out of the calm waters of the channel, I saw looks of concern on a few of the family faces. We arrived at the dive site in no time and begin to suit up. The director and I went over the shots he wanted, which included the ceremony itself, and then shots of the newlyweds riding off into the sunset underwater on their Tekna scooter.

It was rocking and rolling on the boat and I knew where this was going — chum city. I advised the couple I would meet them on the bottom and suggested we swim away from the dive boat a short distance to avoid filming the puke floating down from above, which was going to happen... and it did. I hit the bottom near the reef. It was calmer here than at the sea's surface, but not by much. On the bottom, sea fans were waving around, and the patches of detached seagrass were moving back and forth, everything was in motion. In a few minutes, the wedding couple and the preacher joined me on the bottom and we started the filming. The couple and the preacher were good divers, it was a cool-looking scene. We pulled off the ceremony and went to filming the "ride off into the sunset" scene. I positioned myself in front of this giant Brain Coral with the idea the couple would go flying by with the Brain Coral in the background. The groom and bride mounted the scooter, and I motioned for them to scoot by me. They were coming at me and I was filming. They were looking at the camera. Wouldn't you know it, that enormous chunk of Brain Coral jumped right out in front of them! They were going at a pretty good clip when WHAM! They ran right into the large coral head. It was so damn funny; I was glad I was underwater so they couldn't hear me laughing. We filmed the scene again, but I enjoyed the first shot the most. Swimming back to the dive boat, I could see what looked like wedding confetti floating in the water, but I knew better. The people on board were plenty happy we were back so that they could head for shore. Another successful video job was "in the can".

I called my answering service to get my messages. I had a message about videotaping a shipwreck in Belize.

Chapter 22
World Treasure Finders

A company known as "World Treasure Finders Inc." out of Orlando contacted me reference filming a wreck or possibly several in Belize. A man by the name of Jack Donovan hired me to fly to Placencia, Belize, where I would meet another man by the name of Ted Barrington. Ted would show me a wreck site that supposedly had bronze guns on it. World Treasure Finders was interested in this particular wreck only, considering it might have bronze guns, which would indicate it was a vessel of some standing. I met briefly with Jack and a few others at their office in Orlando. Donovan was also the president of the World Bass Association, an organization that sponsors bass fishing tournaments. We signed a simple agreement outlining my terms and what we expected of one another. They promised me that Ted Barrington had been notified and would be waiting for my arrival.

I needed some help on this one, so my friend Craig Mest agreed to come along with me. Craig was a commercial diver who I'd known since working for Aqua-Tech. I had quite a bit of gear and it made it easier to split it up between the two of us. We drove to Miami where I left my car at a friend's house and we took a cab to the airport from Miami Beach. We headed for the airport, slowly, very slowly. I looked at Craig and he looked at me. We both were somewhat perturbed and equally mystified. Why is this guy going so slow? He had a strange accent. I asked him where he was from. He replied, "Russia."

I asked him, "How long have you been in the states?" "Two weeks."

So now we knew why he was crawling along at 25 miles per hour on the interstate. We encouraged him to go faster the entire trip to the airport or we would miss our flight. We jumped from the cab as soon it stopped at the airport, not offering any tip except, "Go faster next time." Besides, we had no rubles.

We just barely made it, but soon we were flying over the clear waters of the Caribbean. We were flying TACA Airlines to Belize City, after which we would catch a small plane, flying a regional airline called Maya Airlines to Placencia. TACA was the main airline used by the press to fly into Nicaragua or El Salvador to report on the two wars underway in those countries at the time. The standing joke was that TACA stood for Take A Coffin Along. I liked TACA as they served free champagne, all you can drink. It wasn't a long flight, and in no time we were circling the Belize City airport. Looking down, I see a line of Harrier Jump Jets, and some gun emplacements under camo netting along the runway. Belize and Guatemala have had ongoing border disputes for years. After Belize's independence in 1981, Britain continued to maintain British Forces in Belize to protect the country from Guatemala, composed of an army battalion and RAF Harrier fighter jets.

After clearing customs and immigration with no problem, we asked a cab to take us to a decent but not-pricey hotel. The cabby said he knew just the place. The driver took us to the heart of downtown Belize City... if this shithole has a heart. Belize "City" is a scandalous misuse of the word "city". I will refer to the place as "Belize Shithole" as this more accurately describes this sewer drain on the planet. Standing order for all British army personnel stationed in Belize, who went off base, was that for their own protection; they weren't to go anywhere in groups of less than six.

Belize City is a sprawling shantytown populated mainly by the belligerently aggressive descendants of slaves brought over from Africa to log mahogany in the fetid tropical swamps. The sewer system worked like this; you crapped in the Belize River. They placed most outhouses at the end of docks over the river. The idea being the waste went out with the tide.

We stopped in front of a small hotel, went in, and got a room.

The rooms looked like they had been normal size rooms at one time but now were divided in half to make two instead of one. We dropped our baggage off in the room and went to the small restaurant next to our hotel for a drink and something to eat.

The Belizeans seemed to have lost their sense of who they are or how they should behave, particularly the young ones. Much of the youth had been to the states, spent time in big cities, became involved in drugs and gangs, got in trouble, got deported, and brought all that knowledge to the streets of Belize City. If they weren't drug dealing gang members, they thought they were Rasta. The only thing Rasta about them was that they listened to Bob Marley, smoked dope, and let their hair grow. They were just thugs with dreads.

We sat at our table drinking a Belikin beer, a locally brewed beer. I fondly referred to it as "wolf piss". We watched several white people in straw hats, white long sleeve shirts, and overhauls, loading what looked like tools into the back of a pickup truck. These were Mennonites, a religious community of Christians, of German descent. They have been migrating through the years from Europe, to Canada, Mexico, and on to Belize. Mennonites living in Belize exist apart from the government, with limited technology and surrounded by land found suitable for farming. In Belize, they are known for their woodworking skills and house building. We went for a short walk, which got shorter when we heard the slurs coming from the alleys, "Gringo go home." We ended up back in our small room for the night, dreaming about getting out of this shithole in the morning.

The next morning we were happy to be leaving the city and flying to Placencia on a small regional airline called Maya Airlines. They only fly to destinations in Belize and are consistently late, regardless of the destination. We took off in a small white and green airplane flying along the coast. The coastline spans 240 miles from the north to the south. A few coastal communities are scattered up and down the mainland, and only a few of these towns have attractive beaches. Very few of these little communities hold any attraction for tourists. The water on the other hand was turquoise blue, and you could clearly see everything on the bottom below the water.

The flight was only forty minutes, and we were soon circling

211

a grass field landing strip that served as the airport. We were the only passengers who were getting off at this stop and none boarded for the trip back to Belize City. We unloaded our gear and placed it under the one lone palm tree at the edge of the airstrip. We did not know where to go or what to do next. This certainly wasn't the town. There were a few houses close by to the strip, so I walked over to one and knocked on the door. I saw the curtains part and open and close in the window as someone looked out, but the door never opened. I walked back to the palm tree and sat down under it with all our gear surrounding us on the ground. We sat there for a while trying to figure out what our next move might be when I saw a figure across the field standing next to a little creek. I headed over to talk to this person and see if he could help us. He was fussing around with an outboard motor on a dugout canoe. I asked him how I might get to the village of Placencia. He

The author with his gear resting beneath a palm... the only shade at the Placencia landing strip. Photo: Craig Mest

said no problem, he would be happy to take us to the village. We ran back for our baggage and gear, loading it all in the small but sufficient dugout canoe. The dugout canoe followed the little winding creek, flying through the water at a good clip, and we found ourselves in Placencia in short order.

A Belize "taxi"

Placencia has a long history of occupation, starting with the Maya, who established at least fourteen habitation sites around the Placencia Lagoon. In the seventeenth century, the English Puritans who were originally from Nova Scotia and the island of Providencia settled Placencia. This settlement died out during the Central American wars of independence in the 1820s. The Spaniards that traveled the southern coast of Belize gave Placencia its name. At that time Placencia was called Placentia, with the point being called Punta Placencia or Pleasant Point. In the early 1900s the Leslie family, originally from Roatan, also came to Placencia.

One of the other parties involved with this shipwreck venture was Moses Leslie, who is a descendant of the original Leslie Family. Placencia in the 1980s was nothing like the Placencia you would visit today. There was a wide concrete sidewalk that runs the length of the village, a small hotel, a store or two, and a few rum shacks. It also had a movie theater, a large screen TV, and a VCR in a little dark shack that would cost you 25 cents for the matinee. We walked to the hotel and asked about a guest named Ted Barrington, who they confirmed was indeed a guest at the hotel. They said they would summon him for me as he was in his room resting. We sat outside at a table and soon a man appeared identifying himself as Ted Barrington and asked us what we wanted. Ted was of medium build, with dirty blond hair and a three-day-old beard... he looked like he just woke up. I told him who I was, who sent me, and why I was sent. He looked at me. He was waking up from his nap now.

"Wheres my money," he asked.

"I know nothing about any money," I replied. He flies into a tirade about those sons-a-bitches in Florida and he would not show me shit until he got his money. I explained I was only there to confirm the wreck with the bronze guns, and I suspected no money would come his way until he showed me evidence of those guns. I was tired and had come too far to listen to him rant and rave at me. I told him his bitch wasn't with us and we'd be on our way tomorrow if he would not show us anything.

Craig and I got a room at the same hotel as Ted, and he told us to make sure we had a fan to keep the mosquitoes off of us at night.

Ted Barrington on the left, and Moses Leslie

He finally blew off enough steam, and he realized we didn't have any money for him so he started to come around a little. We started drinking, joining the British Defense Force troops who were on R&R at Placencia. These guys were an obnoxious bunch, loud, drunk, and arrogant. The country had several thousand British troops in it at the time. The British had even brought their own radio station with them for the troops in the jungle. It was surreal driving through the jungle surrounded by nothing but wilderness and hearing the *Clash* playing crystal clear over the radio. After several rounds of drinks, Ted finally agreed to show us some shipwrecks, but not the one with the bronze cannon. We lugged our baggage up a flight of skinny stairs to our small room. It had plenty of windows to let in the breeze, but no screens. The sun was going down, we grabbed some fried fish for dinner, and a few Belikans to wash it down. We retired to our room, thankful for the fan. We lay on our beds and sweated, listening to the British soldiers carousing for much of the night. We finally drifted off, lulled by that humming fan.

The next morning we woke up early and went for a walk to check out the area. The peninsula is a 16-mile long strip of sand and sun bordered on one side by the Caribbean Sea and the other by a mangrove-lined lagoon. There would not be a road to Placencia constructed for many years yet. What existed was the sidewalk, the village's only road until 1984. Without a road connecting to the Southern Highway, Placencia was only accessible by sea prior to 1984. Until then, the sidewalk assisted villagers with transporting and delivering essential mainland goods gathered from boats. An urban legend eventually crept into every piece of Placencia travel literature as fact, calling the sidewalk the "Guinness Record Holder of the Narrowest Main Street in the World." Before the sidewalk's two renovations, villagers referred to it as the "Jealous Man Road." Today a man can finally walk comfortably beside his lady if he chooses and no longer has to feel like a creepy bodyguard behind her.

While we waited for Ted to wake up, we had a Belizean seaweed shake. The seaweed shake is seaweed, condensed milk, vanilla, nutmeg, water, and cinnamon. In the evening you can add rum for a cocktail, and it's a great drink. Presently Ted appeared with another

Above: the main "street" of Placencia

Below: a beachside residence on Placencia

man named Moses Leslie. Moses was a large man with a dark complexion, and a head full of dark hair. He spoke in a quiet, somewhat

A view along the shore of Placencia

soft manner with a deep voice. I asked if I could film them while they told me their story or stories. They had no problem with me filming, so I set the camera on a tripod and started my interview with them. We sat under the beautiful palm trees that lined the shoreline, while they told me their story. Moses and his family have been fishing the local waters of the Belize Barrier reef just miles offshore for close to a hundred years. His family also owned the hotel where we were staying and several other businesses in the area.

The stories were like most treasure stories. The once-lost ship was found by accident, unknown cargo, unknown country of origin, but surely carrying treasure. I still have the tape of Moses and Ted sitting under those killer palm trees that are waving in the wind as they talk. The next day they said they would show me some shipwrecks, but not the one with the bronze guns.

The following morning we loaded all our dive gear, cameras, water housings, and the metal detector in a 24-foot inboard/outboard cabin cruiser and set off to the south. I don't know for sure, but I believe we headed south of Ranguana Caye, somewhere between it and the Sapodilla Cayes. After about a two-hour boat ride, the boat

Above and facing page: the author shooting video on the ballast pile of the unidentifed wreck. Photos: Craig Mest

slowed, and we could now see the bottom and the reef all around us. Our boat crept along while Moses got his bearings. Ted jumped up on the bow and threw the anchor as soon as Moses told him to do so. The sun was beating down on us, and I was sweating and eager to get in the water. I put my flippers on, placed my mask on my face, stuck the snorkel in my mouth, and slid into the warm clear waters. I looked down and was greeted by the site of a large ballast pile. The ballast pile was maybe 80 to 90 feet in length, and as much as 50 feet to 60 feet across. It rose from the bottom maybe 12 to 15 feet. I slowly swam around the ballast pile on the surface of the water. I noticed

what appeared to be two cannons fairly close to one another on top of the ballast. They were 7. 7 feet long and maybe a half-foot in diameter. They looked more like cargo than once-mounted guns. Close to the two cannons was an anchor 10 feet long, which was also on top of the ballast pile. It too looked like cargo, or perhaps had been in the hold. At one end of the ballast pile, what must have been the bow, lay two more anchors, ring to ring. These anchors were 10 and a half feet long, the anchor flukes were six and a half feet long. Pottery fragments covered the ballast pile, and over the years the marine growth had sealed everything together. You couldn't pull one ballast stone from the wreck, it was like the entire ballast pile had been glued together. I have only ever seen two wrecks where the ballast looked glued together, indicating that the wreck was untouched by human hands. They were the *Pinta* site and this wreck on the Belize barrier Reef.

I swam back to the boat and put the camera and the housing together. Craig handed me the camera, and I swam off once again, surveying the wreck from the surface above. The sun was now directly overhead, and the rays were hitting the surface and breaking apart into thin columns of light shining down on the wreck. The wreck site

219

was beautiful. It had an immense piece of brain coral incorporated into the ballast pile towards the bow section. A large staghorn coral took its place atop of the wreck, almost like a flag. Towards the aft end of the wreck were several more large coral formations. The wreck had become part of the environment. Nothing looked out of place or foreign, it looked like it was supposed to be this way from the beginning.

Everyone was in the water now, swimming around the wreck, each examining a distinct part of the ancient ship below us. We had a small handheld metal detector that we took turns swimming around the perimeter of the wreck. The amount of pottery on the wreck was amazing; it was fused in with the ballast. There was almost as much pottery as there was ballast. Once I was satisfied that I had shot enough video, I surfaced and took a slate and a tape measure below. Craig held one end of the tape while I got as many measurements as I could before we had to leave. I sketched the wreck, taking notes and measurements as I went. Craig took a few still photographs with one of my cameras while I was videotaping. We had tried to cover as much as we could in the short time we were at the site.

I formulated some theories about this wreck. Was she a merchantman of some type, evident by way of the considerable quantity of pottery on her? Her few cannon would support this idea, as well as her size and location. She sank intact, with no scatter that we could find outside the main pile. Moses stated he had found a large copper pot filled with musket balls. He also said he found grapeshot in a large amphora on the wreck. Was she involved in piracy? All of these possibilities swirled in my head as we headed the little boat back to the mainland about two hours away.

We had been underway for about an hour when someone said, "What's that sound?" The sound was coming from the area of the engine hatch. It was a loud tapping sound... we determined that came from the small inboard motor. We slowed to a stop. Now our only means of propulsion was the wind turning us around and floating us back south towards Honduras. We popped the cover on the engine, Ted stuck his head in the compartment to see what was going on with the motor. I popped open the cooler to see what we had remaining for water and soda, taking a count in case we had to ration them later. Ted and Moses had the valve covers off the motor and were busy trying to fix the problem. Craig had gotten way too much sun that day and once he had no breeze blowing across his face he began looking queasy. We had a radio station from Honduras on, and I figured if we didn't get the boat running soon, the reception was going to get much better as we floated south towards the signal. Ted was able to figure out it was a bad valve, his solution was to wire it so it wouldn't drop, blowing up the motor. After an hour of floating towards Honduras, the sound of the motor suddenly restarting was a welcome relief. We arrived back at the dock in Placencia just as the sun was setting.

The next day Ted and Moses had arranged for us to borrow a different boat to use for the day. It was a 21-foot, center-console, open fisherman, with a nice-sized outboard attached... a relatively new outboard. Yeah! Today they would show us two new sites. We headed in a general due east direction for some time until we could see the reef almost extending above the surface on the horizon. The water went from deep to shallow in the blink of an eye. There was no warning, no scattered reefs to indicate what the waters might be hiding ahead. We donned our mask, fins, and snorkel and swam over the super shallow reef. Our stomachs would almost drag the bottom of the reef in some areas. What lay on top of the reef and all around us were musket barrels. They were scattered everywhere, hundreds if not more, lay on top of the reef. The wood and other small metal parts were long gone, now only the heavy barrels of the guns remained. The reef was covered in scattered wreckage of many vessels. Hull sheathing, fasteners, rigging, spikes, and other unidentifiable iron objects littered the entire reef. It was so shallow I had trouble submerging the camera housing to photograph and video the scene before us.

At my request, we spent much more time here than we should have. Storm clouds appeared on the horizon to our west; a small system was descending from the Maya mountains to our west. The clouds to our west continued to gather, so we pulled anchor and headed to a little cay close by to wait out the coming storm. The local fishermen used this little cay as a camp when far offshore from the mainland. We pulled up to the little cay, which had plenty of fisher-

The author, on the left, is seen here in the company of local fishermen cleaning conch, surrounded by piles of empty conch shells. Photo: Craig Mest

men on it, some cleaning their catch, some making repairs on their boats or gear. There was a small group of two or three cleaning conch. "How much for the conch?" I asked. They gave me four conchs for a Belize dollar, so fifty cents US. They handed me the conchs and were watching me with a look of amused anticipation, wondering what my next move might be. They were thinking, "Let's watch what this white boy does with these conchs." I could see it on their faces. I took out my dive knife, found my spot on the crown, made my hole in the shell, inserted my knife, made my cut and the conch fell out of his shell. I did this three more times, then cleaned the conch. Holding them by their foot I had a nice enormous piece of white meat which I chomped on with gusto as it had been a while since I'd had some conch. The fishermen looked on in amazement and started laughing. They hadn't seen this coming. They flooded me with questions about where I learned to knock a conch out of his shell and clean him? I told them that in Florida, we all knew how to do it, which was a lie, as my skill came from living in The Bahamas for a few years. But I had to stick up for our state, and the old "conchs"[1]. The clouds soon passed over us, and we headed to the west, bouncing over the water back to Placencia.

We stayed another day or two, enjoying the skinny beach and the local rum. It became apparent that Ted was reluctant to show us any more shipwrecks due to broken boats and our lack of any seed money. He said he had business in Belize City and was leaving the next day. We said OK, we'll join you on the flight to Belize City. We found our way back to the little grass airfield and waited on our flight. The little green and white plane landed, and we were once again flying over the clear waters of the Belize coastline. Once we arrived in Belize City we loaded our gear into a waiting cab and headed to the hotel. Ted suggested we stay at the Bellevue Hotel, which is where he stayed when in Belize City. The Bellevue Hotel was situated on the Belize City harbor. They originally built it as a family home in the early 1900s, the site being selected for its proximity to shops, government offices, and other fine residences along with direct access to the outlying cays. We checked in and it was obvious that Ted stayed here often as most of the staff knew him by name. We climbed the stairs behind the bellman who was struggling with our luggage but refused any help. Ted was close behind us as we opened the door to our room and dropped our luggage. We were in the process of tipping the bellman, when Ted, who was headed to his room told us that if we needed anything, the bellman could hook us up. Craig was standing there and asked me, "What did he mean by that?"

I told Craig it most likely meant that if you wanted some weed or hookers, this guy could get them for you.

Craig thought I was joking and quipped, "Sure send a couple up."

The bellman heard him and descended the stairs on a mission. I looked at Craig and told him, "You just ordered yourself two hookers, buddy."

Craig looked at me and said, "Really?"

---

1   Residents of the Florida Keys refer to themselves as "conchs". A saltwater conch is a native, while a freshwater conch is a resident who was not born in the Keys.

"Yes, really you did," I replied. "Now I'm taking a quick shower and getting out of here before your girls arrive and then I'm going to check out the hotel's Maya bar." I showered and got dressed as fast as I could, but not fast enough. The door to the room swung open just as I was putting on my shoes. Two Spanish girls, most likely Guatemalans, with big smiles on their faces, walked in like they owned the place. One immediately sat next to me and the other settled in next to Craig. "Well buddy, you are on your own, see ya, I'm going to the bar," I said as I headed to the door. Craig pleaded with me not to leave him. I knew he didn't speak a word of Spanish, so I thought to myself that might be interesting. I was tempted to let him clean up his own mess but relented and sat back down next to my "date". It was obvious the girls spoke no English. I looked at the girl who was snuggling up to Craig and I mentioned she was much uglier than the one sitting next to me. I further stated that I'm sure mine could most likely eat corn through a picket fence. Craig was becoming more uncomfortable, and I was having fun with his poor decision. I told Craig he was going to need to pay these girls, regardless. Craig said he didn't care, he just wanted this mistake he made to go away. I explained to the girls in my best Spanish that we appreciated them, but that sex wasn't necessary. This statement surprised the girls and hurt their feelings.

One of them said, "Oh, first time?"

I said, "No we are just too tired."

Craig gave them each a twenty-dollar bill and out the door, they walked, confused for sure. A few minutes later there was a knock at the door, and it was the bellman. He was bummed out, "What's wrong, the girls not pretty?"

"They were just OK," I answered.

"No problem," he said, "I'll send for another two girls much prettier."

He turned away on his new mission. I told Craig to get out another twenty and called the bellman back. "Here you go man, we appreciate you but we're just too tired." I lied as I handed him the money.

I finally found my way to the small hotel bar, the "Maya Room". I pulled out one of my Fisheye Productions stickers and stuck it on the wall with a plethora of other random memorabilia. I have often thought about how long that sticker was up on that wall and how many people starred at it, stuck to the wall opposite of where they sat drinking. The hotel has been long gone now for many years, torn down in the late '90s from what I understand. I suppose my Fisheye Productions sticker is buried somewhere with the rubble of the hotel.

I had always wanted to visit the great Mayan city of Tikal in Guatemala. I thought now was the time as we were close, maybe a half day's drive. I inquired at the front desk and they gave me the name and telephone number of a local tour company who would arrange for a car, a driver, and a guide. The price was right, and I told Craig, "We're going to Tikal!"

He asked, "Where?"

I shared my excitement with him that Tikal was one of the largest Mayan cities, which according to UNESCO was inhabited from the 6th Century BC to the 10th Century AD. It was eventually mysteriously abandoned by the Mayans. Tikal was discovered in the Petén Province of Northern Guatemala in 1848. It was opened to the public in 1955 and declared a UNESCO site in 1979. The meaning of "Tikal" is "in the lagoon" but the indigenous people know it as "the place of the spirit voices". The site includes over 3000 structures in its entirety. At least eight separate reservoirs were constructed, and they filtered water with sand brought from some distance away. It's a feat of engineering that enabled the city to thrive as it did for so long.

Our guide arrived with the car and a driver, and we were off to Tikal. Our guide was a young Belizean woman in her early 20s. Her name was Olivia. She was kinda tall, thin, with short hair and a strong Belizean brogue. The driver was a much older man in his late 50s or early 60s. I never heard him utter one word during two days of driving. The car was a black Ford Bronco with plenty of miles on it. I immediately noticed that the windshield would ever so slightly move back and forth in the frame, having been down too many bumpy roads. We left Belize City driving through Hattieville. Hattieville was established

The ruins of Tikal

as a refugee camp after Hurricane Hattie made many people home-
less in Belize City when it hit in 1961, but it became a permanent
town. We drove into Belmopan, the capitol city of Belize. In 1961,

once again after Hurricane Hattie destroyed approximately 75% of the houses and business places in low-lying Belize City, the government built a new capitol city far from destructive coastal waters. This new capitol would be on better terrain, would entail no costly reclamation of land, and would provide for an industrial area. We thought the city looked odd and empty... it didn't seem to blend in with its surroundings. With just over 12,000 inhabitants, Belmopan is the smallest capitol city in the world. We continued driving down the dusty gravel road blowing through tiny villages, leaving a cloud of road dust behind us to coat the fresh

The author stands atop Temple IV as evening descends upon the Guatamalan jungle. Photo: Craig Mest

laundry hanging outside the little thatched grass shacks we passed. I hated this, I thought it was rude, but everyone on that road drove the same way. We drove over the Hawksworth Bridge, which links San Ignacio to the town of Santa Elena. It was a cool bridge, maybe because it's a one-lane suspension bridge, built in 1949 and imported from Middlesbrough, England. On a hill south of town are the Mayan ruins of Cahal Pech, abandoned in the 9th century.

We arrived at the border of Guatemala in the city of Melchor de Mencos, the only major border crossing from Guatemala to Belize. This was a busy place, with children selling greasy bags of freshly fried plantain chips, along with the throngs of money changers running around with fists full of money offering the best exchange rate. On the Guatemala side of the border, there was a gigantic map on a billboard with the words, "Bienvenidos Guatemala" (Welcome to Guatemala). The hilarious thing about this map was that it showed Belize as part of Guatemala. The large map on the billboard omitted Belize but showed Guatemala going all the way to the coast.

We had to get out and go inside along with an assortment of villagers, foreign backpackers, hippies, tourists, business people, adventurers, etc. We got our passports stamped and were soon on our way. It wasn't but just a few miles up the road when we entered our first of many Guatemalan army checkpoints on the road to Tikal.

The Ford Bronco's windshield fascinated me as it went from side to side without collapsing while we bounced down the road to Tikal. We headed north once in the vicinity of Lago Peten Itza, a large lake known for the many Mayan sites surrounding its shores. The road was now even worse, with more military checkpoints to slow us down. Every time we neared a checkpoint, the guide would make sure I wasn't filming, saying over and over out of caution, "Hide the camera."

It was late in the day when we arrived at the Hotel Jaguar Inn Tikal, which was at that time was the only hotel on park property. When we woke in the morning, we heard the sounds of the howler monkeys, in the trees we saw toucans, parrots, and hummingbirds everywhere. We had breakfast and headed into the park to meet up with our Guatemalan park guide, which Olivia had arranged to guide us through the vast ruins.

Tikal was the capitol of a conquest state that became one of the most powerful kingdoms of the ancient Maya. The city has been completely mapped and covered an area greater than 6.2 square miles that included about 3,000 structures. Population estimates for Tikal vary from 10,000 to as high as 90,000 inhabitants. We met Hector, our Guatemalan park guide who spoke perfect English and was an expert on the history and the structures of the Maya. As we walked around the ancient city, we heard only the birds in the trees. We saw no other tourists the two days we were at Tikal. I doubt that's the case now. It surprised me to see the prolific wildlife at the site. There were Spider Monkeys, Howler Monkeys, Toucans, and Green Parrots in many of the trees above us most of the time. I was shooting as much

One of the mighty pyramids of Tikal

video as possible and rarely put the camera down.

My favorite video clip from Tikal is when we were walking a footpath through the jungle with our guides. Suddenly, Hector frantically tells me to drop the camera. Ahead of us was a Guatemalan army unit on patrol walking towards us on the skinny jungle trail. I took the camera off my shoulder but didn't turn it off, holding it down next to my side. As we walked by one another, we exchanged greetings, "Buenos días, Buenos días, Buenos días," (Good morning) over and over as we walked by the soldiers. This is all captured by the running video camera hanging on my side. You see about thirty pairs of boots and fatigues walking past greeting us with "Bueno Dias". It's my favorite brief clip from the Tikal trip.

We spent the day having Tikal all to ourselves with an excellent guide. Hector charged us five US dollars for his services that day, and I haven't found a value like that since. I tipped Hector another five dollars and still felt like it wasn't nearly enough, but Hector was happy. We had explored Tikal with a guide for two days, and I couldn't have been happier with my decision to come see the ancient city. Our last night in Tikal, we climbed the daring ascent to the peak of Temple IV, thought to be the tallest edifice erected by the ancient Maya. Our climb was rewarded with views from the top of the 210-foot-tall structure, which was nothing short of mystical. We could see the Guatemalan jungle all around us, only broken in a few places by the other temples of Tikal, rising above the trees. Smoke hung in the air over the jungle, coming from slash-and-burn farming. An ancient farming technique partially responsible for the fall of the Mayan capitol. It continues into modern times. It was a scene I revisit many times in my dreams.

We left Tikal and headed back to Belize City. It was the same trip, just in reverse. We returned to our hotel and retrieved the rest of our gear. I had a hot cup of tea that didn't seem very hot, and we headed to the airport. We got seated on the plane, and I felt my stomach rumble. The sounds coming from my stomach were ominous. I knew where this was going and prayed for the flight to end soon. That cup of "hot" tea that wasn't quite hot enough had nailed me with foul water. I emerged from the plane with beads of sweat forming on my forehead. I sprinted towards the nearest men's room, unbuckling my belt as I opened the stall door and flew inside. It was as close to a photo finish as I ever want to come. Relief came in waves, not good waves, but things were moving. I'm sitting there thankful for the open stall, and I hear a voice. The voice came from three stalls down.

"Hey, you OK in there?"

"Sorry man, I'm sick," I replied.

"Smells like it."

I took my time, the restroom was now empty except for me. I walked out and headed for the door with my dignity bruised but intact.

I returned to Cocoa and prepared the invoice for World Treasure Finders. I wanted to get paid ASAP; I had taken these people's word more than I should have, and I wanted to get paid. I arranged to deliver my video footage to the World Treasure Finders office near the intersection of Kirkman Road and International Drive in Orlando. Their office was in a two-story building in a business complex. I arrived at their office and was seated with three other gentlemen. I had met only one of them before, and that was Jack Donovan who hired me for the job. They were on speakerphone with someone, making faces, and going through motions like this person is jerking them off. I didn't like what I was seeing with this group so far, and I don't like speakerphones to this day because of this incident. They were soon off the phone and turned their attention to me. They set up a video player, and I showed them the tape.

"Where's the bronze guns?" one of them asked.

"He wouldn't show them to me until you give him the money you promised him," I replied.

"We sent you down there to get video of the bronze guns," another said.

"You sent me down there, knowing this man was expecting me to have the money you promised him," I countered. I can't recall everything said that day, but I finally walked out with my check. I went immediately to the bank and cashed it. When I had their check cashed, I let out a sigh of relief. Something wasn't right here, I thought to myself. They really didn't seem that interested in what I showed them. Nobody said anything much about the beautiful shipwreck I showed them... zero comments. These guys didn't come across as salvors at all to me. I felt lucky I had my money, that was for sure.

Several months later I received a business card from Mike Brick, head of the Orlando FDLE office. There was a note on the card, asking me to call him when I could. I called the number on the card, and he stopped by later that day. When he came to see me we sat outside while he asked me questions about World Treasure Finders. "How long did I know them?" "What did I know about them?" "How often did I work for them?" I shared what I knew and told him about the Belize trip. He seemed more intrigued with the Belize job than he did with World Treasure Finders itself.

The agent was a nice guy. I told him I was suspicious, and I didn't know at the time if I would get paid. He said nothing about why, who, where, or what might be the deal. I knew nothing more when he left than I did when he arrived. Almost a year later to the day headlines in the *Orlando Sentinel* told the story: "Treasure Scam ends in Arrest 3 from Orlando Among 5 Suspects". The story went on, "Five people, including three from Orlando, were charged Thursday with cheating nearly 700 investors out of $1.7 million in a treasure-hunting swindle. Charged with multiple counts of organized fraud, securities fraud, sale of unregistered securities, and violations of the state's Racketeer Influenced, Corrupt Organizations Act were Orlando residents Allen Jones, 41, his wife, Elizabeth, 39, and attorney Ernest Rice, 64; Jack Donovan, 43, Deltona, and Donald Durant, 69, New Orleans." As Forrest Gump would say, "That's about all I got to say bout that."

Chapter 23
Nahoch Nah Chich —
House of the Parrot

In 1987 I attended a Cave Diving Symposium sponsored by NACD and NSS in Tallahassee, Florida. There were many speakers and presentations, but there was one speaker who caught everyone's attention. His name was Mike Madden, originally from Arizona, but now living and running a dive center at a resort called Aventuras Akumal in the Yucatán in Mexico. Akumal is about 60 miles south of Cancun. In the Mayan language, Akumal means "the land of turtles". Mike's presentation was a slide show his buddy Ron Winiker shot, which was of a cave system they called Nahoch Nah Chich. In the Mayan tongue that roughly translates into "House of Parrot". This cave system was nothing like we see here in Florida. The difference being that here in Florida all of our underwater caves are solution caves. These caves are formed by the dissolution of rock along and next to joints (fractures), faults, and layers in the rock. We see no type of speleothems in our underwater caves because the cave would need to be dry for thousands of years to produce any type of calcite formation such as stalactites, stalagmites, flowstone, or columns. The slides that Mike was showing looked like Carlsbad Caverns underwater. Everyone was awestruck by the magnificent calcite formations in this cave. The slides showed every type of speleothem, including "draperies" or curtains which are thin, wavy sheets of calcite hanging downward. There were also immense columns of flowstone. The formations seemed boundless.

Mike stated in his presentation that exploration of this cave was just beginning and that only a few divers had explored the cenote.

Everyone listened intently as the bright, professional slides showcasing the beauty of the system blew us away. When the seminar ended, I waited until Mike was alone and introduced myself. I asked him if anyone had shot any film or video of the cave system, and he said, "No." Which was exactly what I was hoping to hear. I proposed a deal where I would furnish the equipment and compensate him for his time so we could do further exploration of the system. We sketched out an informal agreement and I promised him I would see him in Mexico soon.

I'd just dropped off a new camera at HyperTech in Pompano Beach to have a new housing made for it. That would give me two underwater cameras for production. The camera was the best technology at the time, a VHS-C, smarter, smaller, and easier to handle. I prepared for my first trip to Akumal the best I could considering that cave diving requires a sizable amount of gear. All of your gear is redundant, meaning you have a backup for everything, so you need two of everything. I was taking my cave diving lights plus another large light in a housing with a Gel cell for the camera's illumination.

I finally got a call from HyperTech confirming the new camera housing was ready, so I headed down to pick it up. I enjoyed visiting HyperTech; it was more like a lab than a workshop. It was bright inside, full of acrylic dome ports, stainless steel latches, and young mad scientists hard at work. The new camera and housing was half the size of my other outfit, and I was eager to try it out. Packing for a cave diving video trip out of the country can be a daunting task. I needed to have shipping crates built to protect my delicate technical gear. In the meantime, I would wrap each camera and housing in foam padding and stuff them into large dive bags.

I flew out of Orlando to Cancun, then took a taxi south, past Playa del Carmen to the resort club, Aventuras Akumal. The club was the only hotel on Half-Moon Bay with the best snorkeling, and a killer reef only a few hundred feet off the beach. Some call this stretch of coastline the Riviera Maya. This stretch includes the resort towns of Playa del Carmen, Puerto Aventuras, Akumal, and Tulum, plus the mesmerizing Zona Maya and its unique history. Just two hours south of the glitzy hotels and shops of Cancún, lies the Zona Maya, a stretch of lush jungle dotted with the ruins of ancient Mayan temples. Jaguars and pumas roam the forests, and scarlet macaws flit across the sky. The Zona Maya is also home to communities whose languages and traditions predate colonial contact. Zona Maya is also the central area of the cenotes. A cenote is a natural pit, or sinkhole, resulting from the collapse of limestone bedrock that exposes groundwater. The regional term is specifically associated with the Yucatán Peninsula of Mexico, where cenotes were commonly used for water supplies by the ancient Maya, and occasionally for sacrificial offerings. Here in Florida, we refer to them as sinkholes and springs.

How did the cenotes come to be? Approximately 66 million years ago, a large asteroid crashed into the Earth along the northern coastline of the Yucatán peninsula and into the Gulf of Mexico. The Chicxulub crater, named after the Yucatán town near the center of the crater, is 93 miles (150 km) in diameter, reaching a depth of 12 miles (20 km) down into the continental crust. This massive impact caused earthquakes, volcanic eruptions, mega-tsunamis, and the sky filled with debris and dust for years. There are over 6000 cenotes in the Yucatán forming a ring around the impact zone. It is believed that water from the Gulf entered the limestone bedrock via the impact, and the falling debris collapsed the land's surface into open pits or sinkholes. This cataclysmic impact is associated with the mass extinction event of the dinosaurs and seventy-five percent of life on the surface of planet Earth. And the tourists think they are just swimming in some pretty water.

The cenotes of the Yucatán peninsula include a complex maze of underground caves. The universe of the ancient Maya consisted of three realms; sky, earth, and the underworld. They relate each realm to particular states of being. In the sky or 'kan' live the celestial deities. The Earth is for the living and Xibalba is the watery underworld of danger, deities, and is where life began. Caves and cenotes are the revered entrances to the underworld. Pyramid and temple complexes throughout Mesoamerica are often near these entrances. Underwater cave exploration in the Yucatán has discovered many precious Mayan artifacts and human remains that are believed to have been sacrificial offerings to the gods of the underworld. The shadowy depths inside a cenote are certainly something to be revered. I could hardly wait to get wet in some overhead Mayan environment.

In the 1980s Club Aventuras Akumal was the only resort in that area of the coast with most being farther north. The beach was lined with beautiful, majestic palm trees which have died out because of lethal bronzing. Florida now shares this disease with Mexico, thanks to Hurricane Wilma in 2005, which tracked from the Yucatan to Florida carrying infected treehoppers across the gulf to Tampa. The disease spread when winds blew infected bugs to new territories, or they hitched rides on vehicles. It's said the palm cixiid is particularly attracted to white cars. Back in the '80s, sitting at the tiki bar looking out at the beach, it would have been hard to imagine. CEDAM dive center was Mexico's only NACD (National Association Of Cave Diving) endorsed dive shop. There was acknowledgment but unspoken competition between NSS-CDS (National Speleological Society - Cave Diving Section) and NACD cave divers in the region based on which of them could rack up the newest discoveries and lay the most line. Many new previously unknown systems were being discovered in the Yucatan in the '80s. My heart was, and always will be with NACD, having been certified as cave diver #215 in April 1978 by the organization. I also held a YMCA Cave Diver certification and was certified as a Recovery Specialist by NSS-CDS (National Speleological Society - Cave Diving Section) in 1988.

The newly discovered caves had been given descriptive names like, "The Room of Tears", "The Chamber of the Ancients", "The Temple of Doom", "Mayan Blue", and "Najaron". They named one cenote "Car Wash", simply because people would park their cars next to the cenote and use buckets of freshwater from it to wash their cars.

The tiki bar was a popular place filled with ex-pats, US and European tourists, a few leftover hippies from who knows where. Always fun, most times interesting, and the only watering hole on the beach, my new home. I was looking forward to swimming in some fluids, not just drinking them. I was eager to dive in a Mexican overhead environment. Mike's dive shop was busy, running many ocean dives throughout the day on the magnificent reef just offshore. Also, there was a wall just a short swim offshore that plummeted straight down well past 200 feet. The next day Mike would start showing me around the different systems, so I headed to my room to unplug myself and plug in my camera gear so we could both start fully charged at sunup.

Highway 307 is the main highway running along the coastal area. The buses fly down this road at high speeds, leaving a trail of dust in their wake. The road is paved, with an elevated roadbed with ditches that dropped straight down about ten feet. It's hard to get lost in this area as there are very few roads. The buses drive up and down the highway, taking tourists to and from the Archeological Zone of Tulum.

Our first dive, a popular one that is easily accessed, was the aforementioned Cenote Car Wash, which is situated a short distance from 307. You park close to the water, get geared up, and jump from the jetty into the cenote. The beginning of the cave system is big! The cavern contained locations featuring the charcoal remains of ancient campfires and pottery shards that were clearly visible, left by ancient peoples who used the cave as a dwelling in prehistoric times before it flooded. A tangle of trees crisscrosses the eerie cavern entrance where, over time, they had fallen into the cenote. Mike tied off our guideline, and we entered the darkness. We tied off to the permanent guideline where we tied off our primary reel. Negotiating some minor restrictions, we swam our way back to "Crystal Palace" where I moved past legions of brilliant white stalactites, soda straws, and curtains of flowstone. When we arrived at the "Room of Tears," the cave opened into a large chamber that resembled an ancient cathedral with sculpted columns. Everywhere you turned it was like a cathedral, ornamentation in every direction. I hadn't brought my camera for this first dive; this was more of a "checkout" dive than anything else. Mike needed to know my skill level as a cave diver, and I appreciated that. Safety, knowing your limits, and teamwork are vital. The clock ran quickly and we turned around, based on the air management "rule of thirds" (use one-third of your air going in, one-third going out, and surface with one-third left in your tank).

In some places, the cave system runs close to the surface and picks up brown, tannin-soaked jungle waters. These dark, low viz waters can stain entire cave systems with dirty, brown glazing and obscure spectacular formations. Marine life is conspicuously absent from most cave dives; however, there's usually a thriving marine ecosystem in the open bowl of the cenote. As we floated on the surface at Car Wash, schools of tiny Yucatan tetras would swim around our heads. The beauty of this underwater cave that lay right next to the road amazed me.

Our next dive was Cenote Calavera, also known as Temple of Doom. We headed out of town on the road to Coba, then we headed west on the Coba Road. A few minutes later we pulled off onto a side road where there's a gravel parking lot close by. After parking, it's a small trek through the jungle to reach this sinkhole. The cenote is a large hole, with limestone walls surrounding it, and a drop of 30 feet to the water below. The water depth averages 50 feet in most parts of the system. Floating at the north side of the cenote and looking up towards the entrances, there are two little openings that look like the two eyes and the mouth of a skull. Calavera means "skull" in Spanish, hence the name for the cenote. Here you can also see the color of the water change, with the freshwater being greener and the saltwater crystal clear. There is a halocline in this cave system, where salt water and fresh water meet, creating a weird visual experience where the two waters intermingle. Where they meet it is all blurry, until you are totally submerged in one or the other, then it's crystal clear. When a diver passes through it everything is just slightly out of focus. I refer to this effect as, "stepping through the looking glass". I know of only one similar in Florida; it's in Lake George where there is a spring at the bottom of the lake called "Croaker Hole". In Croaker Hole, trapped saltwater in the bedrock from millions of years ago continues to vent from a fissure in one section of the cave.

Getting in and out of the water at the Temple of Doom was tricky in those days. Mike's tenders would lower our tanks and gear down to us in the water and we donned our gear while in the water. The same routine, in reverse at exit; we would tie ropes on our gear and the tenders would haul them to the top. We jumped in the water and were at the top of a large, circular breakdown pile. A line leads from the top of this pile that runs the circumference of its base. From this line, you have the option of jumping off to either of two downstream cave entrances. We entered via "The Canyons". Another choice could be to go through the "Madonna Passage". These two entrance ways join up in the "Coliseum Room" and proceed onward to the "Hall of Giants". From here you can continue on the mainline, jump off to the "Old Florida Room" or take any of several offshoot tunnels. Maximum depths are around 60 feet, but much of the cave is shallower. We found portions of the cave both above, and below the halocline.

These were the early days of underwater cave exploration along the Rivera Maya, and many of the remote cenotes were unexplored, subsequently being difficult to access. The cave divers were a select few, many from Florida, where most of the underwater caves in the US exist. It was like it was in Florida, back in the '70s… just close the gate behind you, help the landowner police the property, and you could dive just about anywhere. It is much different today in Mexico as it is in Florida. Most cenotes are attractions of some type. Here in Florida, they are campgrounds or state parks. Today in Mexico, they are now attractions where tourists snorkel in the crystal clear waters of the daylight-lit caverns. There is an admission fee to swim, snorkel, or dive. You can read about many of them on Trip Advisor. We can't imagine how the world would change in our lifetime.

I mentioned earlier that finding a new, longer, prettier, cave was a quest shared by many cave divers in the region. You could then name the underground rooms, tunnels, and passageways anything you wished. Naming the unique features of the cave seemed like a big part of the fun. The caves in Mexico along the Rivera Maya differed from Florida caves in a variety of ways. Most have little or no flow, so it makes it easier to swim the passages. In Florida, where many of the caves are first-magnitude springs, you are constantly swimming against a powerful flow of water. The caves in Mexico run shallower than Florida caves. Florida caves can go hundreds of feet deep, whereas many of the Mexican cenotes never exceed 50 or 60 feet deep. Florida caves are water-filled solution caves, water is always present from top to bottom. In Mexico, the caves were solution caves, but they were dry caves for millions of years. As a result, the calcite-laced water seeped from above, creating all the beautiful calcite formations of stalagmites, stalactites, flowstone, ribbon stone, etc. After the asteroid hit, the caves were submerged. In several of the cenotes, there are air pockets. We've taken a sandwich with us on the dive; inflate your BC (buoyancy compensator), come to the surface of the air pocket, lay back, and have lunch.

The story that Mike shares about the discovery of Nahoch is interesting. One of his Mayan employees at his dive shop realized the interest that Mike had in cenotes. The employee recalled visiting another Mayan family with his father when he was young. They lived several kilometers back in the jungle from the main road. This family lived above a large cenote which they used for their freshwater supply, for washing, bathing, food preparation, etc. The employee eventually introduced Mike to the family who owned the cenote and the surrounding property. Mike soon became a friend and a constant visitor to this particular family.

One of Mike Madden's first descriptions of the newly found cave system, published in the 1989 April issue of the NACD Journal, reads as follows:

"When combining the size and beauty of this cave, the net result is miles of ornately decorated passages, giant mounds of snowy white silt, and flowstone formations that appear to be carved from marble. Nahoch's cavern zone has the clearest water imaginable. The stalactites are actively forming in the air space while underwater their color changes from the familiar brownish yellow to brilliant white. Amazingly, the air space runs 500 feet back from the entry area allowing even the inexperienced snorkeler to view the surrealistic beauty of the cave system. Cave diving is swimming through enormous rooms.... some 300 feet wide and a thousand feet long, with columns, stalactites, stalagmites, flowstones, rimstone pools, and cave pearls of pure white limestone and azure blue passages leading off in all directions. Since the system is so shallow, 20 to 30 feet, it requires no decompression. The mainline is still going 9700 feet back in the borehole tunnel, 25 feet in depth."

It was dawn and the dawn chorus had just begun; a clamor of birdsong and the raucous rattle-like calls of chachalacas, parrots, and green jays. Flocks of bright blue Yucatan jays flew through the jungle as we slowly walked alongside the pack horses. The rocky trail was uneven and it was difficult to get a solid footing. We had loaded the two small horses down with four sets of double-104 steel tanks. The horses also carried all our other dive gear and my camera gear. It would take us close to two hours to walk along the rocky path for 2 kilometers before reaching the newly discovered "Nahoch Nah Chich" cave system. The jungle was still cool that morning as we walked into the little dirt farmyard, which sat a short distance from the cave opening. Smoke from a cooking fire floated through the air, roosters were running around chasing the hens. Mike said good morning to the family members in Mayan, which is what most families in this area spoke. They watched us as we walked down a small path toward

The author, on the right, is seen here loading up a pack-horse with dive gear. Photo: Mike Madden

the entrance to the cave where we entered a muddy depression.

Our tenders started lowering our gear down to us, a few pieces at a time. We walked over the muddy floor of the depression on a small wooden walkway to a small platform. The cavern opening was huge, and the cave ceiling was high above the water level going hundreds of feet back from the entry before it slowly tapered off and submerged in the darkness. I loaded the two housings with the video cameras, checked all my settings, and attached the lights to the cameras, getting them ready for the dive. Once the cameras were ready, we suited up. We wiggled into the double-104-cubic-foot steel tanks positioned behind us by our tenders. The tenders helped us position the extensive amount of heavy gear on our backs.

They checked our straps, handed us our flippers, and acted as our butlers, so to speak. Our cave diving friend extraordinaire, Parker Turner, cleverly called this "gentlemen's diving", because you just sat there and let the tenders dress you in your gear.

I slid off into the water, and the tenders handed the cameras down to us in the water... one to me and the other to Mike. I floated on the surface, slowly swimming into the cavern zone. Hundreds of bats hung on the ceiling and the cave walls around us. As I mentioned earlier, the air space in the cavern ran an amazing 500 feet back from the opening. I floated on my back, looking up and around at the immense ceiling above me and its inhabitants, the bats. The ceiling and the water eventually met, I dumped my BC and trimmed my buoyancy before I proceeded much further. As we slowly submerged into the darkness, we flipped on our super-bright cave diving lights. The lights first lit up a large mound of snowy white silt that lay on the bottom. I shined my light up and around the enormous cavern. I was awe-struck. I was in a large room that was decorated with every imaginable type of speleothem (more commonly known as "cave formations") on the planet. From the ceiling hung complex clusters of ceiling decorations called Chandeliers. Soda straws also shared the ceiling. They are slender white tubes, dripping calcite-loaded water out their ends. When their ends become clogged they become stalagmites. Soda straws are some of the most fragile of speleothems. Flowstone cascaded down the walls, looking like undulating sheets of

Because of the remote location of the Nahoch Nah Chich cave system, dive gear was carried in on horseback. Photo: Mike Madden

*Mike Madden gives instructions to his tending crew for a trek into the
wilderness surrounding Nahoch Nah Chich. Photo: Mike Madden*

white marble. The walls were covered with Draperies or curtains
which are thin, wavy sheets of calcite hanging downward, so delicate
they look like fine silk. Stone waterfall formations simulated frozen
cascades flowing from the ceiling down to the floors. It seemed to be
unending! The further we went into the cave, the more beautiful it became. I can understand how caves and their beauty lure unsuspecting
open water divers to their death; it's like a siren's song; "Come, come
see the wonders I have for you, just a little further."

I turned on the powerful video lights and they revealed even
more splendor. It didn't really matter which way you went or how far
you went, you were met with beauty at every turn. Our air supply
slowly dwindled and forced us to end our diving for the day. Once
back to the surface, we dropped our gear to be hauled back to the
ground level by the tenders. Once our heavy gear was strapped to
the backs of the pack horses, we headed up the rocky trail out of the
jungle for the two-hour walk to the main road. It had been a great
introduction to Mike's Nahoch cave!

Mike had a dive shop to run, so I took a day for some sight-
seeing around the Rivera Maya. I visited Coba, which is home to the
tallest pyramid in the Yucatan Peninsula. Measuring over 130 feet

high and 124 steps, Nohuch Mul ("large mound" in Mayan) rewards the climber with one of the most breathtaking panoramas of Mexico.

That evening we prepared to dive into a cave called, "Maya Blue", just five minutes west of Tulum on highway 307. Cenote Escondido (Hidden Cenote) is close to the entrance of Cenote Cristal. In the 1970s, it became known as Maya Blue, when the first SCUBA diver to enter the cave, as is customary, gave it that name. To get to Maya Blue, we had to walk in the dark for one kilometer through the jungle filled with chit palms, many jungle fauna, and other exotic trees. Walking in the dark, loaded down with double steel tanks, lights, and cameras was an adventure in and of itself. That night I almost fell and would have certainly broken my leg had I not caught myself in the nick of time. I was stepping over a tree limb when my footing slipped on the muddy jungle floor. I felt my leg bend backward, almost snapping before I could catch myself and avoid falling. I strained my leg pretty good but we continued with the dive.

The cenote itself is a large, deep, clear swimming pool complete with solid rock cliffs up to four meters high. Cenote Mayan Blue is actually an L-shaped lagoon. Depths differ throughout the system and can approach 80 feet in the saltwater zone. In the freshwater layer, the cave walls will tend to be dark — especially as you get further upstream. The walls of the saltwater layer, in contrast, will tend to be very white. The main tunnel is in freshwater, where the dark cave walls are prevalent. Deeper on, when you jump to the side tunnels is where you encounter the saltwater. After the halocline, at around 18 meters, is where you can see that unbelievable, intensely dark, and shiny blue glow that only exists in fairy tales or maybe in outer space. It's like no other shade of blue, hence the name, "Maya Blue". The cave is not highly decorated, especially on the salt water side, but it has the combination of dark/white, which in this case is not white but blue, a blue that you can't find anywhere else. We used up our air supply and struggled in the dark getting out of the cenote. Finally emerging, we reentered the noisy, lively, night-time jungle for the hike back to our car.

We made several more dives in NaHoch during this trip. However, I wanted to get the footage I had shot back to the states and plan for a return trip to shoot more footage ASAP.

I must mention the nice coral wall, just a short swim from the beach at the resort. We had just gotten back from the long, sweaty walk at the Nahoch site.

Mike asked me, "Want to go see the doctor?"

I answered, "You bet!"

Mike was suggesting that we do a bounce-dive off the wall to 160 feet for a small dose of Nitrogen Narcosis (rapture of the deep). We commonly call this "going to go see the doc" (Doctor Narcosis.). So we grabbed some fresh air tanks, then descended down the wall until our ears started ringing, usually around 160 feet down, then slowly head back to the surface. It was like a "nitrogen narcosis happy hour". We didn't stay down long enough to warrant decompression, just long enough to get a buzz. It was always refreshing.

It hadn't taken long for me to fall in love with underwater caves in this area, and the beaches were beautiful. I thanked Mike Madden and told him I would return soon with more gear to continue filming as much of Nahoch as possible.

When my plane landed in Orlando I went straight to the post-production video studio I used to review the tape. This place had small editing suites where you can view, correct, and edit your video productions. The video monitors face out so that others can see what you're working on, or you can close the door if you wish for complete privacy. I almost always left the door open to get fresh air. As I was viewing the tapes others would walk by and suddenly stop, asking, "What is that?"

"It's an underwater cave in Mexico," I replied.

Soon others had gathered outside my editing suite, watching the tape in amazement. It wasn't long before four or five onlookers joined Mike and me swimming through the cave on one of our cave dives. The video quality was excellent (at that time), but I wanted to shoot some stills and some analog film on my next visit. Film, I felt, would render the colors and the contrast better than video.

Planning and preparations began almost immediately for another trip to the Yucatan. I stopped by and saw my old buddy Craig

Mest, who had offered to build me shipping crates for my cameras and housings. I wasn't back in town long before the phone rang. My ailing father was in the hospital, and they didn't expect that he would live long. He passed away a few days later following a long illness, and the family was grateful he wouldn't suffer anymore. Father's passing set me back a couple of months, but I continued to plan for another trip. Craig Mest had built some nice professional-looking shipping crates for me. Remember this, as it will be very important later.

As I was making my travel arrangements, I realized that perhaps my mother would enjoy getting away. I called and asked her, "Hey Mom, you want to go to Mexico with me?" It surprised me when she said she would love to go. I thought to myself, "Well, this was going to be a different trip than what I'm accustomed to." And I wasn't wrong.

For my second trip to film Nahoch, I wanted to make sure I had all the bases covered; film, video, and stills. I packed two video cameras with housings, a 35 mm Minolta SLR for still images, a 35 mm Nikonos for underwater still images, and I rented a 16 mm Bolex to shoot some movie film. In addition, there were extra lenses and a new underwater light for the second video camera. I filled up the new crates which Craig had made for me with camera gear. Airline tickets were purchased, and I made reservations at a resort close to Akumal. The resort at Aventuras was booked full. I loaded my gear and Mom and was off to the airport. Mom was looking forward to visiting Mexico as it would be her first time.

Our flight was smooth, and looking down at the azure-colored seas from the plane windows was fun for us both. Once we landed in Cancun, we headed through the bustling crowd to baggage claim. I had two full crates and two stuffed bags of dive gear, which I gathered for our trip through Mexican customs. We were directed to Mexican customs, also called "Aduana" for inspection. I placed my crates and bags in front of me for inspection. A customs officer instructed me to open the crates, which I did. As soon as I opened the first crate, the customs officer let out a long whistle. The kind of whistle, which implied, "Wow, look at all the goodies!" They opened the second crate, another long whistle, and another customs officer came over for a look. I turned to my Mom and asked her to take a chair as this could take a while. On previous trips, I had encountered no trouble with customs. I suspected the crates had done me in. They made me look too professional.

The customs officers were saying, "You are a professional, you must have a film permit."

I countered, "No, not professional, Pro-Am (professional-amateur). No permit needed, no money being made."

Not convinced, we continued talking. They informed me I could arrange for a permit. It could take weeks and cost up to $1600. We went back and forth for some time. The best they could do was to allow me to take "one camera, and one housing". They would hold the rest of my equipment at the airport, which I could pick up when I departed the country. This was terrible! I just had thousands of dollars in camera gear seized by Mexican customs. There was nothing I could do but collect what they allowed me to have and walk out the door.

My mother was bubbling over with questions, "Do they always do that?" she asked.

"No, not usually Mom," I replied, trying to remain cool.

I had no trouble when my gear looked like kindergartners packed it! Once I stepped up my appearance, I certainly got attention. Not the kind you want.

We took a cab to the resort, which was beautiful, surrounded by palm trees, blue water, and white sand. This trip had just gone down the toilet, but the smile on my Mom's face and her excited anticipation made it a little easier. I called Mike and told him the bad news, said Mom and I were getting some dinner, and that I would see him tomorrow. Mom and I sat down at the restaurant for a bite to eat.

"What would you like, Mom," I asked.

"Oh, a hamburger would be fine," she said.

So I ordered her a hamburger, and a local catch for myself.

The next morning Mike arrived early, and we planned for a dive at Nohoch the following day. We made several dives in Nahoch over the next few days. I did some more filming the first few days, but after that, we decided we explore a new passage and lay some new lines in the system. We took a day off and I spent it with Mom, who was distraught that the humidity of the tropics was taking the curl out of her perm. I hated to see her upset, so assured her I could fix her hair with no problem. I sat her down in a chair outside, gathered my tools and dippity-doo, and went to work. Out of the many afternoons I spent on the Rivera Maya, that afternoon I will always remember fondly. Mom sitting in the chair, me trying not to make her hair look like RuPaul did it, and us just talking. The breeze blowing through the palm trees, the workers raking the sand, and the tourists walking by. It ended up being a special day. I actually did a good job on her hair and she seemed happy again.

It was time for lunch.

"Mom, what would you like to eat?"

"Oh, I'll just have a hamburger," she replied.

Mom had been eating nothing but hamburgers for days, afraid to stray too far away from food she was familiar with.

I said, "Let me order for you Mom."

I ordered her some "Pollo Frito", knowing that she loved fried chicken. She wanted to know what I had ordered for her, but I told her to wait and see. This made her a little uncomfortable, as she knew I wasn't above a good joke. When they placed that plate of fried chicken down in front of her, she was so happy and surprised.

"I didn't know they had fried chicken in Mexico," she said. "Yes, Mom, they do."

I ordered for mom from then on. She had eaten enough hamburgers for one trip.

During the following days, we explored more of Nahoch. Many of the passages and features of the cave system didn't have names yet, so it's hard for me to describe with much detail or accuracy. My last dive in Nahoch Nah Chich with Mike Madden must have been in a down-stream tunnel, as near as I can remember. Mike was in the lead with the reel laying the line. We were exploring a new tunnel. It was a large tunnel, maybe 40 to 50 feet in diameter. It had curtains of roots in the tunnel which had penetrated the ceiling from above. They hung down like hairy curtains on both sides of the tunnel. I remember how they swayed in the current. We were swimming with the flow, and it was easy to cover lots of ground. New cave passages make you feel very special. You are seeing something that no human has ever seen before. You are the very first person, EVER, to see what is unfolding before your eyes. Just about everything, we saw that day, no other human had ever laid eyes on before... we were the first.

That last dive was a long dive, and we covered lots of virgin passage that day. As we continued down the passage, it surprised us to see a bright column of light shining down from above in the tunnel ahead of us. Mike swam closer. There was an opening in the ceiling. I swam up to the opening. It was big enough for us to drop our tanks and crawl out of the cave to the fresh air above. We emerged from the darkness into the brightly lit jungle. We didn't know where we were exactly, but Mike would soon find out in the days that followed. This was a big deal! Finding another opening to a system can expedite discovering new ground. A new starting point, a new entry point, saving much valuable time. We were both thrilled with our discovery that day.

Mike knew I was leaving the next day. He shared my concern that my equipment was still at the airport. I hoped. Before we re-entered the cave for the swim back, I asked him, "Mike, promise me you won't explore any more of this passage until I can return." Mike agreed, and we named the hole "Mikes Promise". That hole is still known today as "Mikes Promise". We started our swim back to our original entry point, and once again I noticed the roots flowing back and forth in the water indicating a reasonable current. Noticing that the flow on the return trip was naturally stronger, I got a little concerned. We looked at our air supply; we had plenty. I enjoyed the trip back to our entry point, burning everything I saw into my brain. When I crawled out of Nahoch for the last time that day, I did not know Nahoch Nay Chich eventually proved to be the longest underwater cave system ever mapped in the world. And all the credit goes to Mike Madden and his team.

The following day Mike drove us to the airport. I asked him to go in with us in case I needed his help to translate, as his Spanish was much better than mine. We unloaded our baggage and headed towards the customs counter. I didn't see any of the same officers on duty that day that were present when I had first arrived. I inquired about my crates and told them they promised me they would be at the airport when I departed the country. An agent disappeared to a back office, "Whew, they got them," I thought.

He soon returned. "They are not here", he said. "Go see this person," giving us a name on a piece of paper. I was becoming very uncomfortable now. Mike could see my anxiety level going up and told me to just follow his lead and go with the flow.

He then asked me, "How much money do you have?"

"Close to five hundred."

"Get ready," he said.

This was the beginning of going from one person to another, paying each one off to get to the next person. After much negotiating and bribing a half-dozen people, we followed a car to a large warehouse. We entered with a customs officer to try to locate my crates. Looking around, our eyes becoming accustomed to the darkness, we could see lots of stuff. Cars, small boats, outboards, satellite dishes, large screen TVs, motorcycles, crates, boxes, etc. What was disturbing was the fact that we could see many of the seized items had parts and pieces missing off of them. They were slowly being cannibalized for parts. We walked around for a minute when I spied both of my crates. I immediately took inventory. Everything was there, nothing touched.

They escorted us back to the airport where my poor mother had been patiently waiting for us to return. They wouldn't officially re-lease them to me until I had plane tickets out of the country in hand. We had missed our original flight, and there was only one more flight to the US that day. It was going to Houston.

"Come on Mom, we're going to Houston," I said.

"Oh, I've never been to Houston," she replied.

"We won't be there long, hopefully," I responded.

We settled into our seats. I looked out the window to make sure my crates got loaded, which they did. I thought to myself, "Now I can finally relax. Things were getting back to normal." Or so I thought.

The flight was brief, and it wasn't long before we touched down in Houston. We headed to US customs and immigration. Mom and I stood there, as they asked us questions and went through my gear. They were curious why we had bought one-way tickets in cash at the last moment. While I was giving them an explanation, they asked my mother to accompany them to a back room. Yep, they took my mom

The author and his mother in Mexico

to a back room and shook her down. She eventually emerged with a perplexed look on her face. Now satisfied that we weren't smugglers, they allowed us to be on our way. We finally entered the terminal and Mom asked, "Do they always do that when you come back?"

"Not normally Mom, this trip has been a little different."

In 1987 Mike Madden of CEDAM International Dive Center established the CEDAM Cave Diving Team principally to conduct annual projects to focus on cave exploration. During the Nohoch 1997 expedition, totally-explored cave passage surpassed the 37-mile mark. In early 2007, Nohoch Nah Chich included 36 cenotes and had a recorded length of 42 miles when it was subsumed into the 8.7- miles-longer Sistema Sac Actun by the Sac Actun Exploration Team. This portion of the system is now called the "Nohoch Nah Chich Historical section", where, with a 235-foot depth, divers reached the greatest depth of the entire system at "The Blue Abyss". Mike Madden organized the 14 years of exploration in the underwater Nohoch Nah Chich cave system which attracted film productions from around the world including the BBC, National Geographic, CBS, and TVE. Mike continues to explore and film to this day. A quick search online will reveal all you wish to know about Nohoch Nah Chich. I feel blessed to be credited as being among some its first explorers. Thank you, Mike.

Chapter 24
The *Regina* and the Great Storm of 1913

For nearly eighty years this had been her last resting place. Her back is broken and the fire in her heart has been cold for decades. She sleeps most of the time now, it's cold on the bottom, but she doesn't mind. The cold had never bothered her much; she was used to it. It is dark and peaceful where she lies. Occasionally she can feel the storms from above, but it's rare, not like 1913. She remembers her glory days running package freight and grain for the Canadian Lake Transportation Co. LTD, Toronto, a sturdy 249-foot ship. She was young and strong, only six years old.

The *Regina*, as they knew her, had not one but two masts with booms which were used to load heavy cargo into her holds through the hatches on deck. As the ship lay on the bottom, the voices of the 23 crewmen who occupied her decks and quarters had gone silent. The young ship felt fortunate to have such an excellent master. She had always felt she was in expert hands under the command of the young Captain Edward H. McConkey. At 34, he was in his first year of command and one of the youngest captains on the Great Lakes. Despite his youth, he had earned a reputation as a reliable, prudent sailor. The young captain's body had stayed with the ship, trapped in the wreckage after he went down with her, never leaving her alone. It wasn't until almost a year later that *Regina* released her captain from his command. His body was discovered near Lexington in 1914. She didn't want to see him leave her alone. The ship misses the bright sunshine on her decks and the waves rolling beneath her. Eighty feet of water now covers her decks, as she lies on her starboard side fairly intact. She felt the damage to her forward and stern cabins when she hit the bottom, so hard it cast cargo out of her hold. Bottles of Champagne, Scotch, hand lotion, medicine, razors, jars of preserves, and other portions of her cargo lie scattered on the bottom of the lake. Undisturbed precious cargo strewn about waiting, waiting for someone or something to reclaim it. But she lies alone on the bottom in the darkness and the cold, waiting, and waiting.

Then one day she sees a small light coming from above, it descends slowly, growing bigger as it approaches her. She can feel human hands on her, hands that caress her bronze bell and wipe away the heavy silt revealing her name, the name given to her by her creators, the name *"REGINA"*. She can sense the human swimming around her, inspecting her, stopping here and there. It was good to feel a human hand again. Finally, finally, she was not alone anymore.

As Wayne Brusate, a 35-year-old commercial diver from Marysville, Michigan entered the water that day, July 1, 1986, he wasn't aware this would be a date with destiny. He had been checking out a new piece of gear, a WesMar side-scan sonar, which he'd been towing across the water, looking for a small tugboat called the *Scout*. Brusate was the owner of Commercial Divers and Marine Services. He grew up watching the large grain and ore carriers on the St. Clair River, which connects Lake Huron with Lake St. Clair and the St. Clair River. Much like Captain Edward H. McConkey, Wayne loved the lakes and made his living on their waters.

The winds were light, the water calm, and the temperature was 78 degrees that day when Wayne slid over the side of his little boat to investigate what the side-scan sonar depicted as a dark shadow on the bottom. He descended slowly, as he made his way toward the bottom in the 15-foot visibility. A large object gradually came into view... it's a ship, a large ship. She is on her side, her two large masts spread out horizontally on the bottom. Wayne had heard about the great storm of 1913 since he was a little boy, and he was also familiar with the story of the *Regina*. Near one of the masts was a large bell. Wayne swam over to it and slowly wiped the heavy silt away revealing the name, *"REGINA"*. He felt a shiver go up his spine, and he could have sworn he felt a tremble from the ship at the same time. Wayne was

The Regina

flooded with mixed emotions; at first ecstatic, he then became sad, and then curious. He swam around to the bow, where he spied the ship's name painted on the hull... *Regina*. It was an eerie, bitter scene for Wayne, there alone on the bottom, with the Regina and the spirits that surrounded her. His air supply was almost gone, but he did not

Wayne Brusate

want to leave her yet, heading back to the surface at the last second with just enough air for the trip to the surface. As he ascends, sunlight welcomes him back to his natural domain. Just before he broke the surface, he hears what he swears sounds like a woman whispering, "A little oxygen-starved maybe?"

Wayne Brusate's discovery was a substantial discovery, and with substantial discoveries come substantial issues. Issues forced him to keep it a secret. He had two associates with him in his boat that day, both sport divers. Their thoughts on the *Regina* did not fall in line with Wayne's intentions. They wanted her preserved as a Marine Sanctuary. The Marine Sanctuary proposal made little sense to Brusate who only wanted to save the cargo. He had no intention to alter the wreck in any way, nor did he propose such. Had the ship become off-limits to Brusate, it would have been picked bare by sport divers who would sell their souvenirs on the black market. As is so often the case, a bitter dispute between the three former comrades was finally settled in court. The two sport divers, who were sheriff deputies, faced off with Wayne before the bench, disputing Wayne's claim to the wreck. They failed in all their appeals. The secret was successfully kept until Brusate was awarded the salvage rights to the cargo by Michigan's Department of Natural Resources in 1987.

The general cargo consisted of "general store" merchandise; wagon wheels, spikes, nails, Champagne, wine, beer, whiskey, canned goods, crates of cans, horseshoes, silverware, pots & pans, bottles of hand lotion, matches, English jams, razors, and much more. What had really caught Brusate's attention were rumors reported by local newspapers at the time of the loss implying the doomed ship carried $86,000 in gold coin. The money was meant either as payroll for workers on the Sault Ste. Marie canal between the lakes, or perhaps as cash used to purchase grain, which the empty ship would pick up and carry to points south.

The story was repeated in 1916 when the *Sarnia Ontario Observer* reported the ship was carrying $86,000 in coins as payment for a prior grain shipment from Port William. As soon as Wayne gained rights to the cargo, he reached out to Shipwreck Consultants Inc. of Charleston, South Carolina, a firm run by Lee Spence. Subsequently, Lee contacted ORCA Industries and Fisher Research Labs, both of which provided specialized salvage equipment. They brought my company, Fisheye Productions, onboard to videotape and photograph the expedition. Wayne Brusate's company, Commercial Divers and Marine Services Inc. of Michigan would do most of the actual diving.

I was initiated into this venture one evening inside an old Michigan barn. Inside the barn, spread out on tables, in boxes, and in barrels were parts and pieces of the *Regina* cargo. There were bottles, and horseshoes... lots of horseshoes, so many Brusate nicknamed the *Regina*, the "Good Luck Wreck". On the floor, there was fence wire, barbed wire, pots, pans, and on the tables were razors, tumblers, and bottles of ketchup.

This was my first meeting with Wayne and some of his crew, and I immediately liked what I saw. Wayne was soft-spoken, intelligent, hardworking, and quiet. He was dedicated to his family, his job, and his friends. Most of the local divers on the project were also members of the St. Clair County Sheriff's Department Dive and Rescue Team, including Wayne himself, who was in charge of the group. His right-hand "man" was a woman by the name of Colette Witherspoon, a former schoolteacher, and an excellent diver. All the divers on the team were excellent, gaining my respect in the first few days of the operation.

There were two women on the dive team, Colette Witherspoon and Susan Steinmetz. Both women performed as well, and in many instances, better than most men I know. Local diver Brent Deverell and I became good friends, sharing one laugh after another on a daily basis. Brent's favorite type of diving was ice diving. We talked about ice diving and cave diving, comparing the similarities and differences between the two venues. David Losinski was an excellent diver, artist, inventor, and one intelligent dude. He designed and built a working submarine in his garage. David was of Polish descent, and if a person made the mistake of telling a Polish joke, David would ask the offender if he or she spoke Polish? Of course, no one did except David, and he would immediately then ask them, "How does it feel to be dumber than a Polock?"

The vessel we used on the expedition was the *Miss Port Sanilac*. Her captain was Captain Harry Hawkins. The *Miss Port Sanilac* is a 45-foot fishing vessel that was built in 1958 at Pigeon, Michigan. Captain Harry was a local, well-seasoned charter captain who ran flawless operations and knew the lake as well as anyone.

We worked for two salvage seasons during 1987 and 1988. In the 1987 expedition, Mark Brakebill, Jay Shastid, Steve Howard, and I were part of a team brought in by Lee Spence. Mark Brakebill was a former navy diver from San Diego who had worked in the Navy's dol-

The Miss Port Sanilac

phin program, training dolphins for military objectives. Jay Shastid was a commercial diver and former Playgirl male centerfold from Charleston, South Carolina. Steve Howard, captain, diver, and long-time associate of Lee Spence, was also from South Carolina. This was my first time in the Great Lakes region, I liked it, but it was different. I wasn't used to looking out the car window on one side and seeing immense blue waters, then corn and cows on the other. If you asked anybody for directions to any particular spot in the state, they would hold up their hand with thumb extended to simulate the outline of Michigan and show you where to go by pointing to places on their hand.

Port Sanilac was a small village, with a population of around 700 people. The local dive shop we used to fill our tanks in the beginning was in the basement of a guy's house. That was a first for me, the nicest basement dive shop I have ever visited to this day. Diving in freshwater was nice because there was no need to rinse your gear at the end of the day. If only the water were warmer... temperatures between 43 and 52 degrees is cold. We each had two suits, a dry suit, and a quarter-inch-thick wetsuit for a backup suit. Getting a puncture in your drysuit was a daily occurrence. If you made it through an entire day with no punctures you did good. When you got even a tiny puncture, the freezing water at a depth of eighty feet would come screaming in your suit; it was like an electric shock when it hit your body.

The wreck is upside down, with her bow pointed to the north. The visibility on the wreck can vary between one and fifty feet. Water temperatures, depending on the time of the year, range from 35 degrees to 65 degrees in summer. The wreck lays beneath the shipping lanes, and large freighters pass over and close by her daily. I soon discovered that when the big freighters pass close by, it sounds like they are directly overhead.

The *Miss Port Sanilac* was used both years. Captain Harry docked the boat at The Waters Edge restaurant in Port Sanilac. The trip from the dock to the wreck site didn't take long, maybe 30 minutes. At the wreck site, there is frequently a heavy current running over the hull. Descending on the wreck for the first time, you are impressed with the overturned hull slowly coming into view, revealing the enormity of the ship. She was 42 feet across her beam, and 249 feet in length. Resting upside down, her keel broken, the fracture midships provides divers access to her hold through an opening in her side. The propeller was one of the first things to come into view. It is huge! Cargo and rigging laid directly opposite of the break in her hull.

Wearing a drysuit, diver Sue Steinmetz displays one of the Orca EDGE dive computers. The Regina carried domestic cargo including razors.

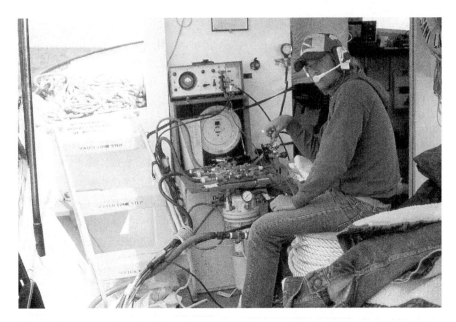

The author is running the dive schedule, the diver communications, and the air box. The air box is a system of manifolds used to furnish surface-supplied air to the diver below. The large guage is a pneumo-fathometer, which is used to determine the diver's depth remotely by measuring back pressure within the pneumo hose.

Her name on her bow and her stern is easily recognized. As one swims toward the large tear in her side, the hold is dark, hosting a tangle of cargo, cables, and rigging. This large tear in her side also allowed entry to the forward area, where we suspected the payroll might be stored.

We stayed within our no-decompression limits to avoid any in-water decompression. The no-decompression limit at 80 feet is 40 minutes, and we usually curtailed it to 35 minutes. To help us gain more bottom time, ORCA Industries provided their new EDGE dive computers for the project, which helped us squeeze all the time we could out of the dives. Rotating five divers, each diver would get in three dives per day on average. We were using Commercial Diving and Marine Services compressors and surface-supplied BandMask gear, along with SCUBA, depending on the task at hand.

I wanted to stay down longer that first dive, but I would soon have all the dive time I wanted. I surfaced and watched the others

Diver Jay Shastid is suited up and ready to go. He is using a Kirby-Morgan BandMask and wearing a bailout bottle. The loose hose on his left side is the pneumo hose. Jay's tender, Mark Brakebill, has his back to the camera.

gear up and go down. Watching the Michigan folks gear up, I could see that they were seasoned divers. They rarely dove in anything but drysuits, and drysuits require extra skill when compared to diving in wetsuits. Drysuits require that the diver wear much more weight than when diving with a wetsuit, I think I had close to 25 pounds of lead on me. Learning to control your buoyancy and venting properly is crucial. I've seen divers blow the fins off of their feet when they turn head-down to descend. If they have too much air in their suit, the air runs up to their feet, and "poof", there goes the flippers. Brusate was certainly among the best, a true professional, always watching out for all the divers. The first day went quickly, and in short order, it was time to become more familiar with the Waters Edge Restaurant and Michigan's premier beer, Strohs, brewed locally in Detroit.

The local divers took over the diving for the 1988 expedition on the shipwreck, which helped reduce operational expenses. Susan Steinmetz and Colette Witherspoon were full-time divers on the expedition, both locals. They also dove for the St. Clair County Sheriff's

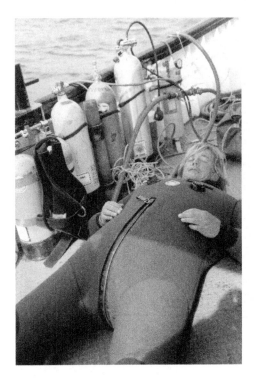

The author inflated this drysuit to test the seals, which apparantly passed with flying colors if this pose is any indication.
Photo: Sue Steinmetz

Search and Rescue Team. Susan and I gave each other a hard time and enjoyed every minute of it. Working with divers like Susan made any problem a small one. Then there was Colette, a giant person disguised as a small school teacher with the heart of a lion. Diver Brent Deverell, was always exclaiming, "A?", as is a customary remark a few miles across the lake in Canada. I picked it up, and it took me months to break the habit. He told me about diving under the ice and the cool stuff they find on the bottom of the smaller inland lakes. And when he says "cool", he meant that much of it was ice tongs and various items that people would lose while harvesting ice from the lakes in the winter. Last but not least was David Losinski who rounded out the Michigan crew. He was an inventor, superb diver, and an artist who produced the first drawing of the *Regina* laying on the bottom. After much work, David made it possible for everyone to see the wreck in its entirety. The dive shop we used most often for both the 1987 and the 1988 expeditions was Rec Diving systems in Royal Oak, Michigan. This shop had it all. It was a large shop, staffed with excellent people. We rented a lot of gear from Rec Diving, especially drysuits. I tried out several brands until I found one I could tolerate.

It was just past sunrise as we lugged the SCUBA tanks down the long dock to the *Miss Port Sanilac*. It took us a couple of trips, two

Using an Aquadyne BandMask, the author is about to go over the side, down to the Regina 80 feet below.

tanks at a time, one bottle in each hand. Captain Harry steered the vintage fishing boat slowly out of the harbor, which was full of more vintage motorboats. As we made our way out of the little marina, I couldn't believe how many boats were constructed in the '40s, '50s, and '60s, all of them in excellent condition. There were antique and classic boats everywhere you looked. No saltwater, no barnacles, no hurricanes, and only used half the year. That makes a difference.

Once out on the lake, huge grain and coal carriers slowly chugged their way past us, sometimes just a few hundred yards away. Hearing the giant ships underwater was enough to make you want to crouch down for cover, even when you're 80 feet deep. The initial work on the wreck required that we clear an area of matches and debris away from the hold so we could penetrate the wreck further. The divers used BandMasks and surface-supplied air with an auxiliary bailout bottle as much as possible. If we were focused on one area and not swimming much, we used BandMask gear. If we had some considerable swimming to do, we used SCUBA. Having a constant source of air is wonderful, the only drawback is pulling around the cumbersome umbilical hose behind you. SCUBA provides more freedom, but without the safety of communications and surface-supplied air.

This area where we found all these boxes of matches was around 10-15 feet high and 30 feet across. A giant mound of wooden matches inside the freighter blocked our way. On this particular day, Wayne, Marc, Jay, and I were on the boat. Everyone had been down and my turn arrived. I landed on the top of the overturned hull and made my way down the hull into the inside of the giant ship. The other divers had done a pretty good job on the pile so far, and I went right to work clearing away my share of wooden stick matches. I worked for a while and was making excellent progress, when suddenly, without warning, the entire side of the match mound collapsed on top of me. It didn't really concern me until I tried to back out of the pile... I was trapped! I wiggled and wormed with no success. I couldn't back out and I couldn't move forward. I started breathing hard and Wayne, who was my tending me, could hear me breathing heavily over the communication set.

"You OK Randy?" Wayne asked.

"Yea, I'm OK, but I'm stuck. I'm going to need some help to get out."

Mark Brakebill suited up and soon appeared out of the mist to help me dig out. It didn't take long and when I was free I gave Marc the thumbs up and we headed for the surface. Once back on the surface, I undressed, removing my BandMask. I should have left it on because as soon as I removed it, the laughs and jeering started. I grabbed Mark and planted a big kiss on the cheek of my hero. They had a good time teasing me about that stupid move. It took days for the ribbing to stop. I even have photos of them wrapping me in line and posing

The Regina carried crates of champagne and good Scotch. Some of the champagne was found still boxed while other bottles littered the lakebed. The diver is Colette Witherspoon. Photos: Wayne Brusate

with me, big grins all around for that one. If they are reading this now, they are laughing their butts off, I guarantee.

The project was covered by a lot of press during the '87 and '88 salvage seasons, both in print and TV. Freedom Marine Inc. out of Vancouver, British Columbia, and Shipwreck Consultants out

Above: Regina carried crates of dishware.
Below: One of the ship's annunciators was found along with the ship's wheel.
Photos: Wayne Brusate

of Charleston, South Carolina, had done a good job of promoting the expedition. They featured Freedom Marine on the Financial News Network during the expedition. Almost daily, press releases carried such headlines as, "Treasure Hunters Race Time", "Secrets of a Shipwreck Being Brought to Surface", "Windy Weather Cancels Dives on Wreck Regina", and "Cold Day on Lake Can't Keep Divers Away from Regina". The local population in the small lakeside communities up and down the shoreline of Lake Huron paid particular attention to our operations. Late in the 1988 salvage season, The Waters Edge Restaurant eventually donated their banquet room for an impromptu museum where many of the artifacts we recovered were on display. One 1988 headline read, "Visitors Come for Regina Salvage." Lee Spence, speaking for Freedom Marine, was quoted as saying that over 1,000 people came on Sunday to see the ar-

The Regina lay on the bottom uncomfortably close to the shipping lanes.

tifacts. Spence said he has had little time to dive the wreck himself because there is so much public interest in it. Press were often present on the *Miss Port Sanilac*, stumbling around and getting in the way. I would spend at least one dive, sometimes two dives a day just video-taping as much as I could, because the news agencies were constantly requesting underwater footage.

The days passed, and the focus turned to the "low hanging fruit" you might say. We were looking at the area where the cargo had scattered outside the ship when she hit bottom and broke open. Wayne and his crew had discovered many single bottles of Scotch and Champagne in this area. The expedition had rented two Tekna DPVs, (diver propulsion vehicles) to help film the wreck. I had used these previously in the Suwanee river looking for artifacts. You weigh yourself down with extra lead, then turn the Tekna's prop toward where you wish to excavate, leaning into it, using the scooter's prop wash to dig. We dug just around the outside of the wreck and im-mediately uncovered more Champagne and Scotch. Isolated bottles dotted the lake floor, eventually leading us to more. With the little Tekna propeller aimed at the bottom, I watched the sand being slowly

After carefully raising a box of Mumm's and Clicquot Champagne to the surface. With less pressure, a few corks flew, allowing for a taste. Photo: Colette Witherspoon

blown away, revealing a complete case of Champagne. The top of the wooden case had the words "Reims France" stenciled and burned into the wood. We carefully blew away the sand from around the case of Mumm's Champagne that turned out to be a full case of twelve and a half bottles. We tied a line to the case of Champagne and carefully escorted it to the surface. At the surface, we handed the case off to our team on deck. The divers scrambled aboard as they prepared to gently remove the top from the case. Everyone gathered around as the lid came off easily and we all gazed at the contents inside. Still packed in yellow straw, with much of the foil around the top intact, lay the bottles, preserved by the cold water and pressure. The bubbly looked like it was in excellent condition. Only the wire that kept the corks in the bottles had rusted away. We pulled out a bottle for closer inspection, when suddenly, "pop". The cork popped right out of the bottle we were holding, then another, "pop". Another cork freed itself from its bottle now that there were no wire retainers and water pressure to keep the corks in the bottles. Somebody said, "what do we do now"? I instantly replied without hesitation, "Get a glass." Someone handed a recently salvaged glass tumbler to me, and we filled it up with 75-year-old Champagne. All eyes were on me, the guinea pig, as I took the first taste. They watched and waited for the verdict, looking at my face for any sign. Was it vinegar or was it good, maybe even better? I handed the glass to Brent, who took a sip. His eyes widened, "That is superb," he said. Now everyone wanted a taste, which wasn't a problem because a few more corks had popped. The Champagne had sweetened but still held its bubbles. We immediately suited back up and escorted the case back to the bottom until we could come up with a solution to the popping corks. The solution was simple, we cut inner tubes into large rubber bands. Once on the bottom, placed the bands on the bottles to keep the corks in them before we brought them to the surface. So we would unpack the bottles on the bottom, place the bands to hold the corks in, repack, and then bring the box to the surface. No more spilled Champagne, but I can assure everyone, not a drop was wasted. Christie's auction house ended up offering 18 bottles of Dewars and Whyte McKay Scotch and 24 bottles of Mumm's and Clicquot Champagne in Chicago. Based on newspaper reports, as many as 30,000 bottles of could be aboard the *Regina*.

We continued to dive day after day, bringing to the surface a variety of items. The state of preservation of many of the artifacts amazed everyone, particularly me being accustomed to diving on saltwater wreck sites. The ship's wooden aft steering wheel was still mounted in perfect condition as if it were waiting for a pair of human hands to steer her one more time. There were bottles of ketchup that still appeared good — it was red and smelled fine — we never tasted it though, that was as far as we went with it. English jams that looked good and smelled fine too. Bowls, silverware, thermometers, canning jars, porcelain pales, ladles, tumblers, razors, and razor blades. The large amounts of alcoholic beverages aboard are attributed to the ports the ship was scheduled to service. Fort William was said to be a colony of heavy drinkers where "water was more expensive than booze". The compass was recovered and saved, still in working condition after 74 years upside down on the bottom. Unfortunately, we discovered the ship's clock and photographed it on the bottom

The ship's compass was recovered and was unique for the candle-house on its side.
A candle illuminated the dial of the compass at night. Photo: Wayne Brusate

before it was prepared for removal to the surface. An approaching storm prevented us from removing the clock that day. When we returned the following day none of the divers could locate the clock and it hasn't been seen since. We salvaged bottles of Hinds Honey & Almond hand lotion which smelled wonderful and everyone took turns rubbing that onto their hands. It looked and smelled like it just came off the shelf. We would joke that if you use this, it would make your hands look 75 years older.

The *Regina* was proving to be a material ledger of her era. Artifacts brought to the surface were carefully logged in a computer and photographed. Weekly reports were submitted to the state of Michigan. The wreck itself was photographed and videotaped extensively. And there remains the mystery of the $80,000 in gold coins said to have been stored in the *Regina's* safe. These coins would be worth more than $2 million today. Some say the money was carried aboard the ship to pay the crew at the end of the season. It was not uncommon for ships in the early 1900s to have large amounts of money aboard. Unlike today's financial settlements, most of the payments for the buying and selling of cargo were done directly via the ship's purser. Purchases were made and paid for at loading docks. This theory would account for the sizable sum of money assumed to be in the ship's safe. But where was it?

As the winds and the waves increased daily, we knew the expedition was near the end of its season. It was just too rough to continue to dive safely. One morning as I listened to the marine forecast on the marine radio the weatherman was calling for 3-to-4-meter seas that day. Converting meters to feet, that's approximately 9-to-13-foot seas, I thought for sure we were landlocked for the day. But the word was that we were headed out momentarily and to get ready. A few people raised their concerns, but nobody wanted to be the one to back out of the dive. Several days prior, I had been videotaping the wreck. As I ascended to the surface, I saw water in the housing's bottom. It was only a few inches of water and had not got to the camera or the camera deck. I had to hold the housing perfectly level as I ascended, so the water would not reach the electronics. As I slowly rose to the surface looking at the water inside the housing, my thoughts were, "Well there goes a couple thousand dollars replacing that housing." So with my main housing down, I spent more time working the wreck than videotaping it.

When we arrived at the site, the seas were rocking and rolling. The waves on the Great Lakes can be hard for any vessel to handle, sometimes the roughest water on the planet. Unlike salt water, fresh water is hard. Rock hard. And those waves can be steep and breaking. That means you take it on the chin, one wave after another in rapid succession. Lake Huron has steep waves, with much less time between them, which is more like running head-on into a series of brick walls. This day was one of those days. The tanks clanged against one another with that familiar tone every diver knows when he or she hears it. Getting dressed for the dive that day was arduous. Just trying to keep your footing while the boat pitched and rocked in the heavy seas was almost impossible. The swim platform at the back of the boat would rise as much as 6 feet into the air with the huge rollers and then plunge back underwater several feet as the wave rolled under the boat. Getting in and out was critical, your timing had to be spot on, or you got hurt. I was eager to get in the water, away from the pitching and rolling of the boat. It would be much calmer on the bottom in 80 feet of water.

I sat on the stern looking down into the water, checked my dive computer, my air supply, and making sure my light was working before I timed my entry between the swells. Splash, over the side I went, headed straight down to the bottom, with the 25-pounds of lead around my waist sinking me toward the bottom. After the many dives I had made on the *Regina*, the sighting of the overturned hull had become familiar. "Plan your dive and dive your plan," is the order of the day. When diving on the *Regina*, every dive can have its anxious moments no matter how experienced the diver. No one ever knows what will be discovered or uncovered.

I landed on the top of the overturned hull, making note of my time, and turning on my high-intensity cave-diving lights. These lights are custom-made, 4000-lumen lights that would light enormous rooms in underwater caves. The light penetrated the greenish, dark waters, illuminating the way forward towards the bow. I slipped into the passage that ran along the side of the ship, and I was thankful to have my powerful light as it pierced the darkness when I entered. This wreck had a sadness to it you could feel. Knowing the history of the ship, having seen pictures of the crew, the young captain, and knowing that the entire crew had perished cast a dark shadow on my psyche that's hard for me to explain.

Now inside the overturned hull, I engaged my cave diving swimming techniques to avoid any contact with anything around me, thereby minimizing any silting, or disturbing anything that might reduce my visibility. I had brought one of my small cave diving reels with me, and I tied the braided nylon line just outside the entrance. The water was clear as I entered a small hallway, tilted on its side, with a few doors on either side. The line rolled off my reel as I eased forward into the now-overhead environment. I could tell I was the first diver to penetrate this part of the wreck, this far into the forward cabin section. Artifacts covered with silt lay here and there, undisturbed. I worked my way slowly along the hall, coming to a door that

I gently pulled open. The wooden paneling that had lined the walls at one time was now collapsed on the floor of the small room. I shined my light into the room, My eyes immediately made out several large lanterns and other material. It came to me I must be in the bosun's locker, where the lamps and navigation lights were strewn about the floor. The silt which had been lying undisturbed, accumulating for many years, now rose from the bottom, as an ominous suspension. I was losing my visibility with each passing second; it was time to go. I grabbed a large oil lamp and then another smaller lamp. Wrapping my cave diving light around my neck, I followed the line out, retracing my steps back to open water. Trying to exit with a large lamp in each hand was difficult, so I bumped along sort of doing a frog hop until I could see the light shining through the opening in the ship's side. I had to set one lamp down while I swam the other to the top of the overturned hull. I left one lamp on top of the hull and swam back down for the other lamp. Once I had both lamps on top of the hull I inflated a little air into my vest and slowly headed towards the daylight on the surface.

As I neared the surface, I could see the *Miss Port Sanilac's* bow rising and then slamming back down into the water. The swim platform was rising violently and then falling back into the water with equal viciousness. The crew had seen my bubbles and were waiting for me when I broke the surface. I swam to the side of the boat as hands reached out to relieve me of the two lamps. I handed them the lamps, happy to have both hands free again I made my way to the stern swim platform. Two crew members were standing on the stern, ready to grab me at the right time and drag me aboard. I stayed back away from the rising and falling platform, watching it go up and come down. I watched the rhythm of the pitching swim platform until I felt I too was in sync. The platform came down, and I scrambled on it as it rose again. Two sets of hands grabbed me and pulled up to the safety of the deck.

I sat on the deck removing my gear as the other crew members examined the brass oil lamps, almost as excited with the find as I was. Now finally shed of my gear, I took a moment to pose with the larger of the two lights.

One of two lamps the author recovered from the confines of the Regina
Photo: Colette Witherspoon

The wind continued to blow, and we cut our dive day short, but I couldn't have been more satisfied with my day's work. My understanding is that one lamp went to the Museum of Arts and History in Port Sanilac and the other to the Dossin Museum of Great Lakes History on Belle Isle.

The salvage project on the *Regina* was not without its crit-

The Regina proved to be the gift that just kept on giving. The author poses here with some of the Dewar's Scotch the ship gave up.
Photo: Colette Witherspoon

ics. A newspaper article headline read, "Current salvage Effort Most Comprehensive of its Kind Involving Lake Huron Shipwreck." This article stated that salvors and Michigan state officials agree it was a model project. But while the salvors say they are doing the state a favor, the state archaeologist says the wreck is better left alone. John Halsey, the Michigan state archaeologist was quoted as saying, "We prefer to leave as much as possible on these wrecks for the benefit of divers. It's much more economical and prudent use of the resource to leave it in place."

"The Regina project is the only major private salvage being conducted in the Michigan Great Lakes," says Sylvan Humphrey, company officer for Freedom Marine LTD of Vancouver, BC who is backing the project. Halsey prefers to see all wrecks left alone so that divers may see the ship and its cargo intact.

"We did not ask somebody to go and salvage that wreck."

Wayne Brusate takes another view. "Anything we bring up becomes property of the state of Michigan," says Brusate. "The state has been taking the best, anything that is historically significant, like the bell,

Diver Colette Witherspoon shows off the bell of the Regina.
Photo: Wayne Brusate

and distributing it to local museums. They've been reasonable. We get to keep the bulk of the cargo. We're able to salvage enough to make it profitable and the states getting valuable artifacts at no cost."

Halsey, the Michigan state archaeologist, when forced, concedes, "It's better to know what's in the wreck, than to have it disappear. The state has neither the money nor the human resources to police every wreck site in the 38,000 square miles of the lake area," he adds.

In the August issue of *Skin Diver* magazine, in the "Wreck Facts" section by Ellsworth Boyd, he wrote, "Michigan diver Randy Lathrop wasted little time rebutting complaints by a diver who claimed the shipwreck Regina - a freighter sunk in Lake Huron in 1913 — was being improperly salvaged. In a letter to Mike Kohut, publisher of "Diving Times" — the paper that printed the diver's letter — Lathrop says: "The structure of the ship has not been altered in any way, shape or form and will not be in the future. Any ship's rigging or compliment has been removed only when requested by the state of Michigan. Not only do divers stand to lose, the public stands to lose if artifacts from this shipwreck and other wrecks are not brought to the surface preserved in museums for the future. Great care has been taken in the treatment of the Regina. It's a model operation; the state and salvors working together so that all people may enjoy the story of the Regina."

John Halsey Michigan state archeologist and James Martindale, chief of law enforcement, Michigan Department of Natural Resources backed Lathrop's sentiments. However, that wasn't the case with Michigan shipwreck preservationist, David Trotter who had his own idea of which he wrote in an article called, "The Future of Great Lakes Shipwrecks — Is There One?" The article parroted the ideas of archeologist-preservationist John Halsey that the wreck *Regina* should be left alone for all to enjoy. Not a single item should be moved from the bottom. He complained about the *Regina's* telegraph being brought up from the wreck. I responded to him in my letter "To believe these wrecks belong only to the diving public is unfair and selfish to all the non-divers young and old. Museums hold many treasures for the public to view and appreciate. The items would be looted one by one. If it were not for the work of salvors, historians, and archaeologists, we would have no heritage to reflect upon if we just let it lay." Trotter also stated in his article, "Perhaps divers, and only divers, who have the most to lose, should determine the shipwreck sites which are to have major items removed." This is such a crazy view, that divers have a "divine right" to make or legislate shipwreck law. Appears this nonsense never ends....

A small portion of *Regina's* Champagne and Scotch was sold by the Chicago firm of Christie Manson and Woods, London, at auction. About a hundred people attended a pre-sale tasting and "were quite pleased," said the auction house. A single bottle of Mumms sold for $90 a half bottle. Dewars Scotch whiskey sold for $70 per quart. "They were bought, mainly by people who we know for fun and their historical value. I know definitely some bottles will be drunk," said Beth Huntman, spokesperson for Christie's Auction House.

The memory of the *Regina* will live a long time. It will hang on people's walls as a spoon, a horseshoe, or one of the many artifacts retrieved and marketed to the public. The viewing of the artifacts at the Waters Edge restaurant brought over a thousand people to the little town on that final Sunday. The divers and crew hung around answering questions, enjoying each other's company for possibly the last time. On the 75th Anniversary of the Great storm of 1913, a commemorative dinner was held. The menu featured, "Captains favorite salad", "November Gale Poached Whitefish", "Low-pressure Champagne Sorbet", "Storm Watch Filet Mignon", "Blowing Tops Rissoli Potatoes", "Force Five Chocolate Mousse", "Whole Gale Roman apple cake", "All Hands Chocolate decadence", and "Snow Squall Cheesecake".

It wasn't easy saying goodbye to the crew and divers. These were people who you had been to sea with, who counted on you, and you had counted on them. We were each other's guardian angels, sharing the darkness, the depth, and the danger that goes with them. We would have put our own lives on the line for one another if that's what it took. Treasure comes in many forms, sometimes it's silver, sometimes it's gold, or it can be other things... the best treasure can be the many friendships you make in the quest. I value those more than any gold. Well, almost.

Chapter 25
The Cape Wreck — A Salvor's Swan Song

She was a Spanish brigantine. They called her, *La Esclavitud*. She was being loaded with supplies and a secret cargo of $40,000 in silver pesos, which is eagerly needed in the debt-ridden city of St. Augustine. It was the spring of 1786, and only two years prior in June of 1784, Governor Zespedes and 500 Spanish soldiers had arrived from Cuba to take over the colony of Florida. In November, the last English refugee ship departed carrying Governor Tonyn, his staff, and the few remaining British subjects away to England. Within weeks, the bustling economic activities of St. Augustine had ended, and the town returned to serving as a remote outpost of the Spanish Empire. The population of St. Augustine decreased from 17,000 to about 3,000, with the majority being the Menorcans. St. Augustine desperately needed the money, which was now, once more, a remote Spanish outpost. The brigantine *La Esclavitud* and two supply ships, one being a schooner, sailed out of the Havana harbor bound for East Florida. On May 3rd both ships were seen off of St. Augustine. As the ships approached St. Augustine, a storm hit, preventing the ships from entering St. Augustine. Five days later on May 8, 1786, the schooner returned to the city without the *La Esclavitud*. The schooner reported that it had lost sight of the *La Esclavitud* near the shoals of Cape Canaveral. *La Esclavitud* was never heard from again.

From November 21 through November 24 of 1984, a low-pressure system formed over the Atlantic Ocean near the southeast Florida coast and developed into an intense storm, unusually strong and well defined for this area and season. The storm produced extensive damage, widespread coastal flooding, and severe beach erosion along the Florida Atlantic coast from West Palm Beach northward to Fernandina Beach. A strong high-pressure area over the northeast US enhanced the severity of the storm. This storm was one of the most damaging storms to affect the eastern coastal sections of Florida during the past several decades. Much of the damage from Fernandina Beach southward to North Miami Beach, a distance of nearly 400 miles, was caused by the easterly winds of gale force with gusts as high as 60 miles per hour blowing for nearly four days. This action of the wind over the ocean produced shore-ward swells around 20-feet high, which pounded Florida's east coast and produced the most severe beach erosion in many areas. To add to the destruction, the highest monthly astronomical tide period coincided with the highest period of storm tides which occurred on the mornings of Thursday and Friday, the 22nd and 23rd. All of this produced tides 4 to 6 feet above mean sea level (MSL) at high tide. In some places, this was the highest tide in the last 30 years. At Mayport, Florida, just north of Jacksonville, the tide of 5.2 feet above MSL was the third-highest tide on record. Much of state road A1A, the famous coastal highway, was closed in Indian River County between Vero Beach and Sebastian Inlet because of high water.

The winds were howling as I tried to open my car door without it being snatched out of my hands by the gale force winds. Once outside the car, the only sound you could hear were the winds shreiking all around you. I had been watching this storm forming for days. It was going to be a doozy, and I was eager to see if the bank of Spain would open for me to take a withdrawal. Storms can be the greatest revealer of riches. It's the storms that sink the ships and drive them to shore, pounding them into pieces, wave after wave, tearing into the ship's structure, ripping it apart, throwing its treasure on the beach and driving them deep into the sand. The storm sank and buried the treasure, and storms possess the key to reveal it again. The winds and waves had been so huge on the beach for the past several days it was impossible to get off the dune, too dangerous.

I walked beaches for a solid week after that storm. Many times I didn't even have a metal detector, because I didn't need one!

8 reales from the wreck of the La Esclavitud.

In the book, "Pieces of Eight", Kip Wagner talks about walking his drunk employee on the beach to sober him up. Then watching this man bend over and pick up cobs from the beach unaided by a metal detector. Well, I'm here to say it happens! I found several coins laying on top of the sand. My eyes would focus on one and then see another a few feet away. I have picked up coins several times since, laying on top of the sand after a storm. After storms I've seen the sand totally washed away, revealing prehistoric fossils laying exposed on the surface. I have pictures of exposed tree trunks hundreds of feet off the beach, revealed at low tide after a storm. I learned a lot from that storm and gained a slew of information which got logged away for future reference. The sands settled back in, and the seas subsided. Everything got back to normal. I burned the memories of walking the beach for days after that storm into my memory. One of those memories was finding the potential resting place of the lost Spanish brigantine *La Esclavitud*, with her cargo of rum, sugar, and $40,000 in silver pesos, now well worth over two million dollars.

The Abandoned Shipwrecks Act of 1988 is a federal law that grants states jurisdiction over abandoned shipwrecks in their territorial waters. The intention of the law was to promote mechanisms at the state level that would protect historic shipwrecks from looters and salvors. One of the most important beneficiaries of this law is the state of Florida, with the longest coastline in the continental United States. This law has been criticized since its inception as it has removed the profit incentive for salvors to discover new shipwrecks. The "Act" has been subjected to a considerable amount of legal criticism because it destroyed the jurisdiction of federal admiralty courts. Bottom line was this: the states were losing ground in their fight against free enterprise. They had to produce new restrictive laws to gain any ground, laws that favored the state, and only the state. It's a bad law and goes against hundreds of years of admiralty process.

The state of Florida claimed title to the wreck of the *Atocha* and forced Mel Fisher's company, Treasure Salvors, Inc., into a contract giving 25% of the found treasure to the state. Mel Fisher and his attorney, David Paul Horan, fought the state, claiming the find should be the company's exclusively. After eight years of litigation, the U. S. Supreme Court ruled in favor of Treasure Salvors and it was awarded rights to all found treasure from the vessel on July 1, 1982. The "in rem Admiralty and Maritime claim" was instrumental in winning this case for Treasure Salvors. I remember David Horan hosting a meeting for salvors at a hotel on US 1 in Melbourne, Florida. Anyone who had a state lease or a potential wreck site was encouraged to file an "in rem Admiralty and Maritime claim" on their site. I want to say this was around 1987 but I'm not sure. However, I remember that the Abandoned Shipwreck Act was on the horizon and this was the last chance to file an admiralty claim and have it grandfathered in before it was too late.

I had shared some secrets of my successful walk with John Brandon, a treasure salvor (Red-beard from Chapter 5). John encouraged me to file a claim on a wreck site we believed could have carried a mint shipment of Carolus III Portrait Dollars dating to the

1780s. I had been hesitant about filing on this wreck because it was in the Canaveral National Seashore. I finally concluded it was now or never. On January 14, 1988, I filed my case, Randy Lathrop (Plaintiff) vs. The Unidentified, Wrecked, and Abandoned Vessel (Defendant). Attorney David Paul Horan filed the motion, and his law firm would represent me moving forward. Almost five months later on June 7th I was awarded Default on the case as no claimants appeared, answered, or otherwise plead to my Complaint within the time required by law. The wreck was mine. It seemed like everything would be fine, that's how little I knew.

On one of my trips to see the guys at the Hyper-Tech camera housing company in Pompano Beach, I saw a boat at a house with a For Sale sign on it. It was a 24-foot Seawind, with a 230 hp OMC inboard/outboard. Basically, it was a small sport fisherman, with a cuddy cabin, and two steering stations, one up top on a flying bridge and one below inside the main cabin. This boat was locally made in South Florida. The owner was fixing her up but had run upon hard times, and with no money, forced to sell her. I could hear the little boat call out to me to rescue her, and so I did. She came with a trailer that I didn't really trust but would help me get her back north. I got her to Fort Pierce and went to work on her.

I named her *Perseverance*. I put new canvas and paint on her, and did she look sweet! *Perseverance* would treat me well and travel to many places in the years to come. We shot fish together, we bugged together, we slept together, we chartered together, we became best friends. Oh, and we hunted for shipwrecks together.

One of her first jobs was performing a 30-day magnetometer survey in July and August 1985. The project was a general survey of the coastline from Ponce Inlet to the North Cape area. I had contacted Kirk Purvis, and he had agreed to help me. We used a Varian Cesium Magnetometer, with Kirk serving as the operator. I was the skipper, and Tom Kraft from my East Coast Research days joined us as first mate. We ran the project out of Ponce Inlet.

When we first arrived with *Perseverance* at the marina in Ponce Inlet, things did not go well at all. Kirk and I were slowly approaching the boat slip, and I climbed up to the top steering station

to see better. What I didn't know but soon found out, was the previous owner of the boat, for whatever reason, had cabled the top controls in reverse order. So forward was reverse, and reverse was forward. Expecting the boat to go in reverse as you threw the gear into "reverse" you would actually put the gear into forward without realizing it. Kirk was below on deck with our docking lines, waiting for me to navigate her gently into our boat slip. I went to throw her in forward and the circus act was on. We started doing loopty-loops in reverse in the marina. This got everyone's attention. Resting feet came down off the rails, cocktails went flying overboard, deck chairs were abandoned. Kirk was down below hanging on for dear life, going, "whoa, whoa, whoa" as we went around in circles. I thankfully got her throttled down and the gears in neutral. My heart was beating, Kirk was down below laughing, the other residents were shaking their heads and giving us the stink eye. We eventually salvaged our reputation with our neighbors, unfortunately, that's all we salvaged this trip.

We got quite a few days pulling that mag down the coast. Often, as soon as we were a mile south of the inlet, we would throw the fish overboard and start magging. I knew more than I knew before, but I did not know enough quite yet.

With the Abandoned Shipwreck Act looming on the horizon, I was eager to get going with a contractor. I had been in contact with a company called Ocean Enterprises, LTD from Lantana, Florida. The person in charge was a man named Paul Kruger. He had two boats, the larger of the two was the *Ocean Explorer*, and the smaller one being the *Monty K*. The *Monty K* was around 48 feet and well equipped for a project like the Cape project.

I had arranged to meet the company's Dive Systems Engineer, Tom McGehearty, in Fort Lauderdale. I flew into Fort Lauderdale and Tom picked me up at the airport. It was just after noon, and Tom asked me if I'd like to go have a donut. I thought to myself, I'd much rather have a beer, but maybe this guy doesn't drink. Tom looked just like Mr. Clean as he suffered from alopecia areata. This means that your immune system mistakenly attacks a part of your body. When you have alopecia areata, cells in your immune system surround and attack your hair follicles. Tom didn't have a hair on his body. So, Tom pulls into the parking lot of the Donut Shop, we get out, and I notice

Above: Kirk Purvis

Below: Tom Kraft

the windows to the donut shop are covered? As we enter, I see a sign that states no one under 18 admitted. We enter the shop and sit down. A waitress soon arrived and asks us what kind of donuts we would like today. The waitress was a nice-looking girl. Just one thing. She was topless. Tom had taken me to Florida's only topless donut shop. He was grinning from ear to ear when he saw my reaction. Thanks to the donut shop, I had renewed hope for "donut-eating" Tom, with no thought of drinking any beer. We finished our donuts and coffee, took one last look at the fancy cream dispensers, and went on our way.

Tom and I jumped in the car and we headed for Paul Kruger's house. He lived on a canal and was swimming down the middle of it when we arrived. Tom struck me as a straight-up guy, but my first impression of Paul was that he was, shall we say, "odd". He was a small wiry guy with glasses that rode down low on his nose. He certainly seemed to have the boat and gear required, so we hashed out a temporary agreement.

Paul and the research vessel *Monty K* arrived at the site on June 23rd, 1988. Onboard Paul was acting as captain, and there was a first mate, a cook, and four divers, plus myself. We docked at Ponce Inlet and continued with more survey work for several weeks, until July 11th, 1988.

It was a Monday and I had stayed ashore because I had a phone conference scheduled with David Horan, our attorney. I'd told Paul to continue with the survey work and not to do anything *but* survey work. Several weeks previous, in my absence, Paul had attempted to come in the back way via Mosquito Lagoon in a Zodiac to get to the beach for a look at the area. He had the divers with him, and it was good he did because he got stuck. They ended up in shallow water and had to push and pull their way out. Paul was a challenge to deal with. I suppose I didn't speak loudly enough that day because Paul did what he wanted to do, anyway. They cruised down to the site and Paul came up with this great idea about sending the divers into the surf zone with metal detectors to have a look around. It didn't matter how stupid this was, or how deep the sand was, or that we had told him before his chances of finding anything were slim to none.

Later that afternoon when I returned, I was briefed on their foray into the surf. Paul had sent three divers into the surf zone with metal detectors. While the three were flipping around in the surf, they'd come up, and stick their heads out of the water to get their bearings. On the beach, they saw three nice US Park Rangers on three-wheelers waving at them. Divers, Curly, Moe, and Larry wondered what could the nice Rangers wanted from them? Maybe asking if they've seen any picnic baskets on the bottom out there. So, they swim to the beach to see how they can be of help. The Rangers were thrilled that the guys had come to accept the tickets the rangers had ready for them. Each diver got a ticket for having a metal detector in the Canaveral National Seashore park... and they were diving with no dive flag. Oh, I almost forgot — all three metal detectors were confiscated. I couldn't believe anyone could be so stupid to swim in to the beach to get busted. I asked them in disbelief, "Why did you swim to the beach when they waved you in?" In addition, they were buzzed by Park Service helicopters during the entire incident.

The authorities were shown the wreck's "arrest warrant" and the federal admiralty papers, but they didn't care. They further advised that the vessel *Monty K* stood a chance of being seized just as the metal detectors had been seized. It would take two months before the metal detectors were returned, but we got the charges dismissed rather quickly. The government's attorneys had all the information they needed and were preparing a preliminary injunction. We waited at the dock until August 19th, waiting for legal relief, but it wasn't coming that summer.

We were trying to get a preliminary injunction ourselves, to protect us from, "The agents, employees, and attorneys of the United States including, but not limited to, personnel of the Canaveral National Seashore and the National Park Service", from interfering in our operations or my case, against the wreck itself. Hearing nothing about my injunction from our attorneys, we continued with our survey work on the site. I wrote several letters to the Park Service advising them of our pending survey work, listing vessels and crew members but never heard from them or received a reply of any kind. Fortunately for me, I had other work at the time in the Great Lakes. I wrote the Park Service on August 15th, 1989 notifying them I would be back on my wreck site on August 20th, 1989. This made it easy for them to prepare to harass us. I had hoped for a different reaction and continued to try to open a line of communication with them. When they would fly low over us in the helicopters, they never answered us on Channel 16, so it wasn't a surprise. I had hired James Sinclair, a marine archaeologist, to plan an archaeological design to ensure the archaeological provenance of the wreck site area would be maintained.

The *Monty K* arrived at the site on August 20th and continued with remote sensing survey work. We had recorded a decent number of anomalies during our survey and were hoping the injunction would soon be granted so we could start digging on the site. A tropical wave was moving off the coast of Africa into the Atlantic Ocean. The National Hurricane Center upgraded the depression to Tropical Storm Gabrielle on August 31, forcing us to pull off the site for safe harbor. Five to thirteen-foot waves were reported along the east coast of the United States from Florida to Maryland. We were in the heart of our hurricane season, the Atlantic had come alive.

Around September 10th, a new storm was rising from a cluster of thunderstorms near Cape Verde. This storm would become known as Hurricane Hugo. Along the coast of South Carolina, Hugo set new records for storm surge heights, reaching over 20 feet near McClellanville, South Carolina. Our salvage season that never really got properly started was over.

Our day in court finally arrived, and we appeared at the Federal Courthouse in Orlando, Florida. Our attorneys, John Brandon, Debbie Brandon, and I sat in the courtroom pews waiting for our ruling. Park Service personnel were also present that day in full uniform. The funny thing was they kept giving John Brandon the stink eye; I believe they thought John was me. They didn't just do this once but filled John full of holes with their dirty looks. I thought it was hilarious and teased John about it. There were other bureaucrats from the state in the room, but they weren't identifying themselves. David Horan did an excellent job of defending our case. Judge Sharp listened intently. When he ruled, he simply stated he was supporting the constitution. The gavel came down, and we had our injunction! Order- THIS CAUSE is before the Court on the Plaintiffs Verified Emergency Motion for Temporary Injunction and The response thereto by The United States Of America. WHEREFORE, it is ordered and ADJUDGED and DECREED as follows: The Plaintiffs Motion For Preliminary Injunction is hereby GRANTED. The federal and state bureaucrats were not happy and filed quietly out of the room, trying not to look at us anymore. They hated the smiles, hugs, and backslapping. Damn near killed them.

I moved on from Ocean Enterprises as my contractor per mutual agreement. I now teamed up with Cobb Coin Company and John Brandon for this late start on the wreck site. We wouldn't have much time, but we had to go for it.

The press release read as follows:

"Mr. Randy Lathrop of Ft. Pierce, Florida, has announced that he is beginning exploration and recovery operations on an unidentified shipwreck he discovered off Cape Canaveral National Seashore in 1984, and it is believed to have sunk in the 1780s.

Mr. Lathrop sought out Admiralty attorneys, Mr. David Paul Horan, and his brother, Mr. Ed Horan, of Key West, Florida, to represent him in his assertion of his federal rights in Admiralty Court. Mr. Lathrop was granted Federal Admiralty rights to the site and appointed substitute custodian by Federal Admiralty Judge Kendall Sharp. On August 6, 1990, Mr. Lathrop was further issued an injunction against interference in his operations by the National Park Service.

Mr. Lathrop has contracted with famed salvor Mel Fisher and his firm, Cobb Coin Co., to do the actual work. Mr. Fisher will use his considerable resources in the form of recovery vessels, divers, archaeologists, historians, preservationists, and laboratory facilities in the exploration and study of the site. Mr. Fisher has dispatched the M/V Endeavor to conduct initial explorations of the site under the direction of Capt. John Brandon, 37, of Ft. Pierce, Florida.

Lathrop, Fisher, and Brandon also stress that they are not seeking an adversarial relationship with either the National Park Service or the State of Florida. The participants in the project feel it is important to the people of Florida and the United States that our maritime heritage be recovered, in an archaeological fashion, preserved, and studied. Many shipwrecks, such as this site, are in high energy zones, shallow water surf zone, and are exhibiting signs of ongoing deterioration, and efforts must be made to save the artifacts and information they hold. Neither the State of Florida, nor the National Park Service has the funding, equipment, personnel, or commitment to undertake such recovery efforts. Mr. Lathrop and Cobb Coin will fulfill this need at no cost to the taxpayer.

It is also pointed out that every Spanish galleon discovered, to date, in Florida waters was researched, located, and recovered by the private sector. All the treasures and artifacts in the state's sizable collection, worth tens of millions of dollars, and priceless in historical terms, are from the private sector. The state has never undertaken to recover historically important artifacts from a Spanish galleon shipwreck in Florida waters. Preferring instead to having a policy of leaving the artifacts on the seabed where they are imperiled by the actions of the sea. Lathrop and crew hope they can add yet another page to the legacy of the private sector recovery of history from beneath Florida waters."

The ruling made national news, paper after paper carried the headlines, "Judge allows Treasure Hunter to Salvage Shipwreck in National Park."

In early August Captain Kim Fisher brought the *M/V Bookmaker* to the dock at the Ponce Inlet marina in company with Captain John Brandon who commanded *M/V Endeavor*. The blowers affixed to both vessels created a stir in the marina. Once we had the boats tied up, I saw John talking to someone who I later learned was Burt Webber, Jr. Webber seemed surprised and confused as to why we were there at the marina[1]. Another man came up to John Brandon, and he was as excited as if Santa Claus had just come down the chimney. His name was Jack Scarborough, he had lived in the area all of his 60 years, and he knew many of the players. Jack was friends with Don Porter, Dan Porter's father, a former treasure salvor himself. He had known many of the local salvors, beachcombers, and treasure hunters. Jack had stories galore, lots of local treasure lore. We would see Jack often in the days that followed. He was as excited as we were. Or maybe more so.

In the days that followed we continued with our remote sensing survey. We magged the entire length of the admiralty arrest area, from a depth of approximately five feet to a depth of twenty-two feet, in sixty-foot passes. We logged over 135 mag readings. Unfortunately,

---

1 Burt D. Webber, Jr. is a celebrated salvor who succeeded in recovering a great deal of treasure from the wreck of the Concepcion lost on Silver Shoals, north of the Dominican Republic in 1978. At the time the author met him in New Smyrna, Webber was involved in a venture to recover some cannons, just south of Ponce Inlet at New Smyrna.

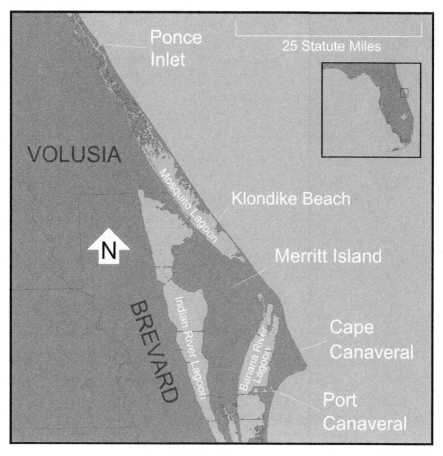

La Esclavitud wrecked about midway between Port Canaveral and Ponce Inlet within the confines of the Canaveral National Seashore.

none were of sufficient size to represent cannons or anchors, with most being quite small. We chose several of the largest hits as excavation targets. Almost as soon as we lowered the blowers, here come the helicopters once again, buzzing us and not answering us on marine channel 16. The helicopter flew low and we could see more park personnel on the beach taking pictures. Now the Park Service and the state were in cahoots with one another to end our efforts any way possible. The state, unbeknownst to me, had filed charges against me on August 8, 1990, just two days after Judge Sharp granted us an injunction! If you need a dirty playbook, ask the state... they have plenty of them.

The civil action by the State of Florida, Department of Natural Resources, Department of Environmental Regulation, and the Division of Historic Resources, Florida Dept of State, all wanted a piece of my, you know what! They had filed four violation complaints against me; they wanted fines of $10,000 per day for every day we were on the wreck. The newspapers had picked up the story from our court appearance and they were running with it. The headline in the *Daytona News-Journal* read, "State Readies Legal Broadside In Dispute Over Treasure Hunt." State Assistant Attorney General Eric Taylor had fired off one attempt for an injunction against the salvage effort, only to have it blocked by my attorneys. Taylor said Monday as he was preparing another salvo of legal maneuvers.

Taylor contends, "This is our land. These are our waters. We have a legal sovereign right to it. He (Lathrop) is digging 20 feet into the sand. This is mining. This is excavation."

"Our position is very clear", Horan countered. "And that is the admiralty law has outright court jurisdiction over navigable waters."

I had the court order dated August 6, from Judge G. Kendall Sharp in Orlando that named Taylor in a footnote and warns against him trying to stop me. The article went on, "Both sides are clearly bent on fighting the case, and neither side had anything complimentary to say about the other."

Taylor and the state's motion to block us in state court was denied. U. S. District Judge G. Kendall Sharp warned Taylor not to stop us from working the wreck. The court order stated that Taylor, "is aware of this court's jurisdiction over (Lathrops) salvage activities. The court is in sincere hopes that the state of Florida is not seeking to attempt..... interference with the plaintiffs (me) activities." The order called the state's efforts, "nothing more than bad faith harassment of the plaintiff." In the newspaper articles, Taylor bragged he had park rangers, Florida Marine Patrol, and rangers on a helicopter monitoring our activities. He also told the *News-Journal*, "the state hasn't ruled out physical intervention by law enforcement personnel if the salvage crew was spotted excavating sand off the site." I thought to myself, this guy must be going for the State of Florida's, "ARRR Matey Award", it's an award the state gives out to state employees who board salvage vessels with a cutlass in their teeth and steal the booty. Just kidding. But they really should have an award like that for themselves. They certainly deserve recognition for the mean work they have done over the years.

I shared my side of the story with the paper, telling the reporter we had sought a working relationship and tried to cooperate with the state and U.S. park officials since we began the project. But received nothing back but harassment. Further, I said, "We have extended invitations to them to have their archaeologist come work with us, which they declined. They wouldn't say why." I was pissed about the way I had been treated by the Park Service and the state. Taylor didn't stop. He told the papers that if we removed anything from the site, we would be arrested. The article ended with me sharing my concern for the safety of the artifacts. I told the reporter, "In hurricanes and storms, artifacts can be destroyed. We're trying to save Florida history."

"Taylor could not be reached for comment Wednesday night," the article concluded. We never heard another threat from Assistant Attorney General Eric Taylor, thanks to David Paul Horan and Judge Sharp. But it wasn't over. I would hear from Eric Taylor again, claiming he wanted to "cut a deal" supposedly, but that was only a stalling tactic. The bureaucrats had simply retreated to the shadows to plan another attack. Another government agency would now come after me. The plan was to wear me down, tag-teaming each other, so when one would fail, another would step in the ring.

We worked every day with no further interference other than getting buzzed by the occasional helicopter. I had always hoped for a quick discovery in the form of cannons or anchors that might show the bulk of the shipwreck. This proved not to be the case. We started digging as many hits as we could. We chose several of the largest hits and several groupings of hits. We chose hits on the south end, in the middle, off where the bulk of the coins were found, and on the north end to excavate. All proved to be material that could not be associated with the wreck. We found a single ship's spike close to the beach that could possibly be related to the wreck, certainly some wreck. We

Captain John Brandon has moved his vessel, the Endeavor, into position and prepares to lower the blowers. Note the tube extensions added to Endeavor's blowers, necessary to dig in the deep sands along the north side of Cape Canaveral.

weren't able to work right on the beach where we needed to excavate because the court order came in so late in the season that the ground swells, which occur in the latter part of the year, had already started to roll, producing breakers in the shallow water, and keeping us further offshore than where we wanted to be. Keeping these points in mind, several possibilities arise. The first might be that the ship had lost its anchors offshore and was armed only with bronze cannon, eliminating any large mag hits. This is probably not the case since most ships of the period were not armed with bronze guns. And anchors are almost always present on these shallow-water sites. A second possibility could be that the ship went down in deeper water directly offshore, and only a small section washed in. The most likely scenario based on the evidence is that the ship wrecked in shallow water either to the north or the south of where the coins come ashore. This would be in keeping with circumstances on many of the 1715 wreck sites, where large amounts of coins and artifacts are found on the beach,

Captain John Brandon, left, and the author considering their next move as the sun beats down on the distant beach of the Canaveral National Seashore.

but the wreck site that produced them might be several miles away.

I suspect that this is the case and that when excavating close to the beach, as we cut across the scatter, following it north or south to the limits of the permitted area. I'd then file for an extension of the boundaries and hopefully follow the scatter to the main shipwreck area. This would require the weather to cooperate with us, and we needed to start the season early to avoid the large swells that customarily rolled through later in the year. We still needed to check more of the mag hits we recorded, as we'd only begun that work. Some of the targets could offer significant clues.

We spent a month at Ponce Inlet, which brought on considerable expense for dockage of our two vessels, motel room bills, and food for the crew members. The boats had to run the 52-miles round-trip every day... that's a lot of fuel. Compare that to most of the 1715 sites that required less than a 10-mile round-trip. Both the *Bookmaker* and the *Endeavor* blew transmissions during this short season, which added unexpected expense. All of this added expense plus the deteriorating weather led us to the logical conclusion that our season was over for the year. We prepared the vessels to head back south to Fort Pierce. But not before Mel and Deo Fisher came to town for a party that Jack Scarborough was hosting for the crew. We had gotten to know Jack well over the past month. He was always eager to help any way he could. More often than not meeting us at the dock every evening as we pulled in, to see if we needed any help carrying the gold and silver ashore. The party was a nice finish to a tough summer. Mel and Deo Fisher were there, John and Debbie Brandon, Jim Sinclair, me, all the *Endeavor* crew, the *Bookmaker* crew, and lots of Jack's friends. One of my favorite photos of all time is Jack smiling from ear to ear, Mel Fisher has one arm around his shoulder and I have my other arm around him from the other side. Everyone had a great time. Someone told the newspaper, and they ran a story on it, which we didn't know until many days after we headed back south. I always tried to keep in touch with Jack after that summer. We would go for car rides around the area. He would show me places and share local lore with me. Jack taught me how to recognize the low-lying midden mounds that are up and down the backside of A1A. I see my surroundings differently now when I drive down the roads; I see things I didn't see before I knew Jack. He joked with me about watching out for the "music worms" in the backwoods, which is what he called rattlesnakes. Sadly, Jack passed away before the next salvage season. We accomplished at least one wonderful thing that summer, and that was to have Jack Scarborough meet his hero Mel Fisher, and that made us all feel good. Rest in Peace Jack.

Now I had time to think about what might happen next. The cost of running such a long distance every day was making the effort more difficult than first envisioned. The expense of litigation was devastating to me. John and I both agreed that the site held good potential, but the expense of maintaining the operation at Ponce Inlet would not be fair to his divers or investors. I was grateful for the splendid effort that John Brandon and Mel Fisher had put forth that summer to help me find my wreck, but I understood why they couldn't come back next season. John still wanted to try to work with the state but I had no desire to have anything to do with them. But I

A grand gathering for Jack Scarborough... from left to right standing in the first row is Jim Cardenas, John Brandon, Deborah Smith, Terrie Thorbjornson, and Jack Scarborough's wife. Sitting is Eric Bush, and on the extreme right is Jim Sinclair. Back row standing from left to right: Deo Fisher, Mel Fisher, Jack Scarborough, Jacques Lemaire, and the author, Randy Lathrop. Photo: Eric Bush

tried; I had applied for a permit with the state in May of 90, which was denied, as I knew it would be.

In December 1990, I heard from Eric Taylor that the state had approved a settlement, in principle to conduct salvage activities. There were two reservations: he declared that additional time was necessary to get final approval and prepare a written agreement. After receiving approval from Florida, I would be required to get permission from the United States. In retrospect, this was probably just a ploy to keep us tied to the dock while they plotted their strategy. I waited for 6 months for the state to process my application to excavate until I couldn't wait any longer. A key requirement of admiralty law is that the salvor put forth as much effort as possible to rescue the cargo of the lost and wrecked ship. This is to ensure the cargo is recovered as soon as possible and to prevent salvors from claiming areas to just keep others away. As a salvor, I had to prove my ability to perform and show that I had a good chance of succeeding. With this very much on my mind, I made a new agreement with David Sands of Sands Salvage at Port Canaveral, Florida. I notified the Park Service as a courtesy and

The MV Dragon II was the last boat to do any excavation work at the La Esclavitud site.

to ensure that additional parties did not try to benefit from the ruling by United States District Court Judge G. Kendall Sharp.

The major salvage vessel that would now be used was the *M/V Dragon II*, 41 feet with brand new twin blowers hailing out of Port Canaveral. In addition, we had the 20-foot *Blue Runner*, and another 20-foot support boat, the *Mullet Choker*. Both smaller boats had blowers that could be attached easily when required. Jim Stiller and his Brother Hubert Stiller were two local net fishermen who would join us occasionally. Both men knew the local waters well and had been fishing them for many years. They knew the fast, simple way of getting to the site was to hug the shoreline along the coast where, right off the beach, there was a ditch if you knew those waters that you could traverse right up the coast to the site in pleasant weather. To get to the wreck site from Port Canaveral, it's necessary to pass by the entire stretch of the Kennedy Space Center, which presents some security issues.

My logbook for July 19, 1991, notes that we left the dock at

0600 hours, arrived on the wreck site at 0720 hours. We noted a 3-foot depth coming over the bar, just enough to float the 40-foot *Dragon II* over the same. Then at 0945, a ranger appears on the beach and takes pictures. These were brand new blowers, it was the first time using them. Unfortunately, one of the blowers wasn't working, and we left the site at 1100 hours. The blower was fixed overnight, but large swells kept us in port for two more days. We went back to the area where we recovered the ship's spike last year. The bottom here was noted as sugar sand. On a normal day, we would try to arrive at the site by 0730, then work until 1300 hours. The southeast winds would start to blow after 12 noon. If the waves increased in size, we couldn't take the short route home because the waves were too big inside the slough next to the beach. I noted in the ship's logbook that the sand was deep, and the holes filled in constantly. You had to maintain a high RPM on your props to keep the hole open. The visibility was awful on the bottom. A diver could see maybe a foot at the most. Here the bottom is all sand, and there is nothing to hang on to. The surf moves you back and forth constantly. I was concerned that we were not getting to the hardpan. The bottom looked like a mixture of gray sugar sand with a mixture of shell fragments. The month of July was hit and miss, but we pulled about a week of solid work out of the month. Park Service helicopters continued their surveillance, making three low passes at us on July 28th as noted in the ship's logbook.

The beginning of August saw the *Dragon II* back in port for additional fabrication on the blowers. I had arranged for Kirk Purvis to come with us and try to relocate several of the mag hits from past surveys. Kirk arrived on August 10th, but the mag malfunctioned and we finally located a technician to fix it on August 15th. Anytime the space shuttle would land at the Cape, we would be shut down.

The blowers were really working well now; we were hitting coquina rock and peat, which was a good sign. It was still slow going, and it would take us sometimes as much as 40 minutes to blow a hole and hit hard bottom. Much of the site is deep sand, as much as 20-feet deep. We would have to blow four holes to make a square, then we would blow a hole right in the middle of those four. This way the sand had some place to go, and we could keep the bottom of the hole cleared. On August 16th we set up on a 90-gamma hit we had we had recorded the year before with John Brandon. We blew for 40 minutes to get to the bottom. The hole was ten feet deep and twenty feet across. The bottom was noted as being shell and peat. At the bottom of the hole was a 3-inch by 2-inch piece of terra cotta pottery found using a metal detector. Some fired pottery will be picked up by a metal detector. The entry for that day also said, "Incredible visibility, the best I've seen in all my years on the site." I also noted, "Optimistic about pottery shard being so close to 90 gamma reading." We also noted a ranger on the beach at 1425 hours. We finished the month feeling like we were making progress.

September kept us at the dock for the first 10 days before the seas would allow us to work. On Sept 13th, 1991, we arrived on site, had the blowers down at 0915, and we're digging hard with bedrock showing now in 15-feet of sand after blowing only 20 minutes at 1100 RPM. Suddenly a different helicopter descended on us, coming out of nowhere and flying low and dangerously close to us. This helicopter was yellow, and it appeared that they were filming us. Then later that afternoon, we spotted two more rangers on the beach at 1420 hours, and one was taking pictures again. The following day, September 14th, we arrived on site around 1000. The ground tackle was down and we were blowing by 1105. We noticed a ranger on the beach and a small boat slowly making her way through the increasing seas. The boat was a small Boston Whaler-type boat, maybe 18-feet long. In the boat, we could make out two figures in uniforms getting jostled around by the swell. The small boat finally came within shouting distance, "Permission to come aboard," one ranger asked.

"Permission denied," I replied.

They then asked me, "Are you the captain?"

"No, my name is Randy Lathrop, what do you want?"

They perked up and replied, "Oh, the man himself!" They handed me a paper. It was a cease and desist order from The Department of the Army Corps of Engineers in Jacksonville, Florida ordering me to stop immediately or criminal penalties may be imposed including up to $2500 in fines and one year in prison. The signatory of the order, Colonel Terrance C. Salt, also threatened me with civil fines of up to $25,000 per day. The two rangers in the boat that day were Mike and Jim, I'll not provide last names so we don't have to put them in the book's index.

To say I'm not happy with this sudden change of events would be an understatement. I looked at the two rangers bouncing around in the little boat and told them, "Whoever sent you down here in that little boat and these rough seas think little of you evidently." I can't remember exactly what more I said, but it didn't stop there. They said that they couldn't leave the site until we did. I told them it was going to take me some time to get all our tackle in so we could leave. We then pulled out our lunch and had a nice leisurely meal while we watched those rangers bounce around, up and down in the increasing seas. The rangers continued to bounce around in their little boat while we worked slowly but surely. Watching them would be my only satisfaction with this turn of events, and I savored every little splash that soaked them more than they already were. We finally hauled all our gear aboard so the now-soaked rangers could retreat to the park and perform their bonafide duties.

As I looked at the cease and desist order and I could see the Department of Defense symbol in the left-hand corner. I'm thinking, "Now I'm fighting the Department of Defense!" Actually, it was the Corp of Engineers, but what does it matter at this point. I had been cited under Section 10 of the Rivers and Harbors Act of 1899, 33 U.S.C. 5403, prohibiting the placing of any structure in or over navigable waters of the United States and excavating from or depositing material into such waters unless the work has been properly authorized by an Army permit.

Now let's think about this charge against me. So now our salvage vessel has become a "structure". How did our boat become a structure? They had to reach back to 1899 to find something to charge us with and shut us down. They had no regard for admiralty law but went back to 1899 to enforce this little gem of a law?

My case now looked like this: *Randy Lathrop-Plaintiff vs. The unidentified, wrecked & Abandoned Vessel Defendant. State of*

Florida, et al., plaintiffs, vs. Randy Lathrop, Defendant. The two cases had been merged in federal court into one case. As the regulator of the Cape Canaveral National Seashore, the United States asserted its paramount role in protecting the land from further excavation. The Corp of Engineers insisted it had jurisdiction over dredging activities occurring in tidal water that extended from the mean high water line to the outer limits of the continental shelf. This case has many elements, and is convoluted, even making it hard for the attorneys to understand. The bottom line was that we followed through with the permitting required by the Corp of Engineers and were denied. The reason the permits were denied was because the State of Florida's Department of Environmental Regulation had denied a similar request. Does any of this sound like it makes any sense?

This controversy presented two complex issues involving the principles of jurisdiction, federalism, and comity, and Congressional power to alter substantive admiralty law, namely; (1) whether the court has "in personam" [2] jurisdiction to issue an injunction against the United States and its agents where the United States is not a party to this litigation and has not been served with process, and (2) whether Congress can constitutionally supplement substantive admiralty law by regulating salvage activities; and if so, whether the United States can require a potential salvor of an alleged historical shipwreck to comply with federal law requiring a permit before conducting salvage activities in a national park. We applied for many permits with the state and federal government but were denied every time. Our next move was filing for a Second Motion for Preliminary and Permanent Injunction, which we did on April 6, 1992.

A year had passed, and the case had been assigned to a different judge, Judge Harvey E. Schlesinger in Jacksonville, Florida. We had heard nothing back from the court. We were no longer confident about the results in this case. Ed Horan had spoken to the court's law clerks, and they led him to believe that the decision would not be in our favor. Ed Horan felt that with the horrible cases that had been handed down recently regarding salvage, he no longer felt it would

---

2   In personam is a Latin phrase meaning "against a particular person".

be worthwhile to appeal a judgment if it went against us. Within the last week, the United States Supreme Court denied Cert on the Central America case. The Appellate Court's ruling which held in favor of the insurance companies and established their right to maintain a claim against treasure that had long been abandoned was just AWFUL. There were seven states that filed Amicus Briefs in support of the salvor's position and the attorneys were sure that the Supreme Court would have to accept the case... they didn't. It was a travesty on the waters of justice... it didn't stop there. The Federal Marine Sanctuary Act that encompasses all the waters surrounding Monroe County (Florida Keys), outlawed salvage, made the backcountry areas off-limits for boating, and could eventually keep the public from doing anything and everything. They have even defined snorkeling as a "consumptive use" over which they have jurisdiction. We are being regulated out of existence, and I'm sure the founding fathers are rolling over in their graves.

Approximately one year after Judge Schlesinger stated he would have a decision on the merits of our case within two to three weeks, we received a decision denying our Motion for Preliminary Injunction. So basically, not only was the judge late on providing a decision, but he didn't attempt to deal with the cross-motions for Summary Judgement upon the facts that were agreed to. Basically, the attorneys felt that the judge had made it very clear as to how he would rule on these matters, and they were extremely upset about the way the case had been handled. The attorneys at the Horan Law firm felt that Judge Schlesinger provided one of the worst opinions they had seen under the law of Admiralty. It was obvious that the good judge and his clerks had little familiarity with Admiralty law and specifically the law of salvage. His starting points were all based on faulty interpretations of the law, and it got worse from there. He basically held that there is no law of salvage unless and until a salvor complies with all existing regulatory schemes whether they be state or federal. Historic precedent was ignored, and the judge had apparently taken away our right to roam the seas in any attempt to locate wrecked or abandoned vessels that are in peril. He also applied the law of finds[3], not as a supplement to Admiralty law, but as a means to an end. He has made the Abandoned Shipwreck act a useless piece of legislation and indicated that the state, as the owner of the submerged land, would have the right to refuse salvage. I could go on and on but I think there is little point in further explanation.

Little did I realize, way back in 1977 when we found the gold tray that it would lead me into a profession so full of excitement, intrigue, danger, fascination, and amazement. On the flip side, I have come to recognize that it also involves hypocrisy, corruption, and a multiplicity of incompetent, uncaring, egregious governmental agencies.

The Cape Canaveral coastline is unique, not because of the geographic boundaries established by law, designating it as the space center or as a national park, but because it has been a predominant landmark noticed by early sailing explorers in their excursions to the New World. As today's population looks to the heavens in amazement and to the wonder of space travel, the eyes of past generations looked upon the coastline with the same amazement and wonder for other equally stimulating reasons. The federal court ruling will only have a bad effect. Shipwrecks will continue to be found. The marine environment will, from time to time, uncover and expose its hidden history to the diving public. With no recourse or legal procedure in place to compensate the honesty of the finder when he comes forward in good faith, much of that which is recovered will be lost to the black market. The actions of the State of Florida have promoted the impression that pirates are no longer identified by eye patches, cutlasses, and a parrot on their shoulder. But rather by their stodgy suits, briefcases, and the government titles they go by. The State of Florida and the National Park Service should incorporate the Jolly Roger into their government seals, so we may be forewarned of whom they really represent.

I had lost years in the pursuit of a dream. I was broke and in debt. My spirit was crushed, my future uncertain. I wanted little to do with any type of salvage or treasure. I just wanted to forget as much

---

3   A salvor who finds a shipwreck pursuant to the *law of finds* is entitled to the full value of all of the goods that are recovered.

as I could and as fast as I could. It wasn't long before I eventually lost just about everything; I had no money and soon lost my house too. I had to get serious, forgetting about salvaging a shipwreck, rather focusing upon the salvage of my life instead. Count me as being done with treasure, or so I thought anyway… treasure would find me the next time.

Somewhere along Florida's coast...

Chapter 26
Hurricane Irma — The Ghost Canoe

She was born as an African Easterly Wave, more commonly known as a tropical wave. Coming off the coast of Africa, she became a tropical storm on August 30th, 2017 and was given the name, "Irma". The storm attained hurricane status on August 31st while still in the far eastern Atlantic about 650 miles west of the Cabo Verde Islands. Irma continued to strengthen over the next several days as she tracked westward across the tropical Atlantic. By the time Hurricane Irma struck Barbuda in the northern Leeward Islands around 2 am on Wednesday, September 6th, she was a Category 5 hurricane with 185 mph winds. On Sunday morning, September 10th, Irma weakened to a Category 4 hurricane as she sped up toward the Florida Keys. The eye made landfall on Cudjoe Key as a 130 mph Category 4 storm at 9:10 am. The center of Irma then proceeded westward and northward making a second Florida landfall in Marco Island at 3:35 pm that afternoon as a Category 3 storm with 115 mph winds. The center moved into central Florida overnight.

I lay in bed listening to the house creak and moan as the winds screamed outside. The house seemed to breathe, as its fastenings and shingles fought to hold on against the hurricane-force winds tearing at it. I had been watching this storm as I do every storm, waiting. Maybe mother nature would force open the Bank of Spain? I had my detectors charged and ready to go. I had just retired, after over 20 years in the trade show business. I produced shows from Seattle to New York. Working with trade unions wasn't always easy and I was happy to spend some time at home for a change.

Why do the strongest winds always come at night, when it's dark, I thought to myself? I hate lying in bed in the dark listening to the sounds outside, the crashing, the howling, the rain hammering on the roof, forced into every little crack by the hurricane-force winds. Neither Suzy, my wife, nor myself could sleep, we just lay in bed listening to the winds and the roof creaking and moaning. Suzy and I said silent prayers to ourselves, praying that the roof would stay on for another hurricane. Both of us have seen many storms in this house, we never evacuate; that's not our style.

Once again we had gotten lucky, the winds for our area were 58 mph sustained; 80 mph max gusts. The morning came, the storm passed, the sun rose, and the clouds raced across the sky at high speed. I walked outside. Tree limbs and debris lay scattered everywhere. Needing fresh air, I hopped on my bike with my GoPro on the handlebars and went to see what had happened on Indian River Drive. Our house is just north of the Beachline (State Road 528) sitting near Indian River Drive, which runs along the Indian River Lagoon. I headed south to Cocoa Village but didn't get far because a large portion of Indian River Drive was gone, just a big hole where the road used to be. Since I couldn't go south to Cocoa Village, I turned around and went north instead. The water was still over the road in many places as I slowly made my way past City Point. This is the area of the old settlement of City Point, settled shortly after the Civil War by Confederate veterans, citrus grove workers, northern winter residents, and "consumptives" seeking a healthy climate. My friend Craig Mest, who went to Belize with me and built my camera crates, owns property and several acres at City Point. On the hillside of his property, there is an old private cemetery. It's a small family cemetery with graves containing confederate civil war veterans. I rode by Craig's house, which seemed to have made it through the storm OK. My GoPro started beeping letting me know I had no more battery remaining... "It must be a bad battery," I thought. I rode on towards the north, trying to avoid the pockets of saltwater still standing on the road in many places. Bicycles and saltwater don't mix, they hate one another.

Debris covered the road, trees, boards, portions of docks, boat cushions, life preservers, all types of flotsam imaginable. The storm had

Irma's winds pushed water from the Indian River Lagoon over its western bank on to Indian River Drive near the author's home where the dugout canoe came to rest.

tossed large coquina rocks that had once lined the shoreline across the road as if they were Styrofoam. I thought to myself that it was unfortunate my GoPro battery had gone bad on me when I had just started my ride, and now I couldn't video the trip. There were two or three large sailboats that were cast upon the shoreline, laying on their sides. I got to the old church on Indian River Drive and turned around. My bike was squeaking as I rode along, surveying the damage along the river bank. The day was gray, and the clouds were moving through the sky at a good clip. I came around a slight bend and saw something that I immediately recognized. It was about 15-feet long, carved out of wood. I had seen these before. I had ridden in several of them in Belize. It was a dugout canoe. I chuckled, "Isn't this the damnedest thing you ever saw," I thought to myself. It was a cool-looking dugout; it had two separate compartments, and the bow looked like a gator head. I took out my iPhone and snapped a few shots of it, one which I immediately shared with Jim Sinclair, my archaeologist friend. I rode just down the road where I saw Craig

The canoe as found on Indian River Drive in Cocoa, Florida.

standing outside his house looking around.

"Hey dude, you gotta see this dugout that got washed up on the road a little further down," I told him.

"Let's go get it," Craig replied.

We jumped in his pickup truck and drove the short distance to where the dugout still laid in the road. Craig dropped the tailgate and we attempted to lift it into the truck's bed. It was much heavier than we thought. We struggled and got the front of it in the truck, then grunted and groaned until we had the whole thing completely loaded. I felt my back give, but my excitement seemed to overshadow the pain. I checked my phone and saw that Jim Sinclair had replied to the photo with a simple, "WTF". Now that we had it secured, we looked it over. It looked like it was possibly cypress, it had square nails in it, and some sections looked painted. Once we got the dugout in the back of the truck, we headed to Craig's house. We had rescued it just in time because as we looked down the road we saw a county crew clearing and picking the debris up from the road. But... damn,

I'd thrown my back out again!

Craig and I unloaded the dugout and started looking at it closely. Craig was almost certain it was cypress. I went home and got my close-up lens and came back to photograph every minor aspect of the dugout. After I was satisfied I had taken enough pictures for the moment I went home. I texted Jim Sinclair and asked if he could get in contact with the State of Florida's BAR (Bureau of Archaeological Research). The storm made it difficult to contact anyone with the state, and Jim knew most of them personally. I dropped my camera bag and sat down and pulled out my phone to look at the first pics I took that showed it sitting in the road. Opening the Facebook app on my phone, I posted the following:

"Look what Irma kicked up out of the bottom of the Indian River, a dugout canoe. Florida State Dept of Historical Research has been notified, it is the law to notify the DHR."

I didn't think too much more about it. I was more concerned with when the power would come back on. I turned off my phone to conserve power and went outside to clean up the debris which was everywhere. It was getting hot outside, so I took a break for a cold drink and checked my phone. I opened up Facebook to see what was going on and checked out my dugout canoe post. I couldn't believe my eyes! The post had over 10,000 shares in less than one day, and about as many "likes"! My post had gone viral! I looked at my message box and it had over 50 messages. The messages were from many media outlets and newspapers. Several local TV reporters wished to talk to me, along with a slew of newspapers and magazines. The Facebook post didn't stop there, it continued to be liked 104K times, with the shares soon exceeding 93,000! The messages kept coming too. I couldn't keep up with them if I had wanted to. I spoke to Jim Sinclair, and after having a good laugh about all the attention, he confirmed he had gotten in touch with the state guys to take this thing off my hands. It wasn't long before I got a message from Julie Duggins, an employee of the BAR (Bureau of Archaeological Research. She started off by saying, "over 25,000 shares, wow." She said that she had contacted a local archaeologist who would come by and take notes and pictures. His name was Tom and he would contact me tomorrow.

Tom showed up the next day. He was irritated that the powers that be had called on him during an emergency to look at this canoe. He wanted to be back in the motel room he and his family had occupied during the storm, not out here in the aftermath of Hurricane Irma. Trying to make small talk with him was fruitless. He said it wasn't his field, he didn't know why they called him, etc. I didn't know and didn't care, and I stopped talking. We took measurements, watched each other sweat, as I listened to him grumble until he was satisfied he had done what was expected of him. Finally, he drove away. Craig and I both agreed that people can't always be what you'd like them to be. Civil maybe? I should have known from past experiences the academics were just getting wound up.

The following day Craig and I slipped the dugout into a freshwater pond behind his house to prevent it from drying out. I had watched Tom, the archaeologist, take pictures with his point-and-shoot camera, and knew he couldn't compete with me as a photographer. I contacted Julie Duggins and asked if she wanted some shots and she said yes, she would like that. I sent quite a few closeups, and she was thrilled with the detail and quality.

As previously mentioned, the day following my discovery of the dugout canoe, I was flooded with requests for interviews from print and television reporters. The Facebook post was still gaining steam, getting thousands of shares and "likes" every day! I did several local TV spots with Channel 6, Channel 13, and Channel 35, along with several live interviews on the phone. The photos of the dugout had gone viral as well. They were being picked up and printed around the world by foreign news agencies. I received calls and messages from Europe, Asia, Russia, and many other places, all wanting to know if they could use the pictures. The truth was, I had lost control of the photos. They were flying over the internet at warp speed. At least I was getting credit for the photos. I had always dreamed of one of my photos going viral, but not one I had taken on my phone. I was getting friend requests and gaining followers on Facebook, sometimes hundreds a day. The politicians were sending me friend requests, and everyone wanted to see the dugout canoe. The news agencies wanted to send photographers... I said, "No need, I'm a professional and I have all the photos you can use." An ad agency that promoted the Space Coast contacted me. The ad agency lady and I became good friends. She would schedule the interviews and press releases for me. The more press that hit the airwaves, the more popular the dugout canoe became. It was crazy.

I had been on the phone with Julie Duggins, the state archaeologist, assuring her that the dugout was submerged in fresh water and safe. Everyone wanted to see it, but I allowed no one to see it. If the press wanted any pictures, I had plenty. I was aware that the only compensation I would receive for all my time and trouble would be the exposure I would get as a photographer. Anyway, there was no reason to disturb the dugout. I had also received several messages on Facebook offering to buy the dugout canoe. I never entertained the idea, so I don't know how much the offers might have been. The press interviews kept on coming, and they occupied my entire day for several weeks.

A wood sample from the dugout canoe had been sent to a lab for radiocarbon dating. When I first saw the dugout canoe, several things struck me; there was a variety of nails spanning a hundred years or more in various places on the dugout. The shape of the bow looked like a gator head or dragon head of sorts. The wood seemed to have been chosen because of a unique knot on it that made it look like a head. There were patches of paint here and there on it, some blue, some red. I felt it wasn't so much of a canoe as it was an outrigger. There were marks on the side that might show they attached it as an outrigger to a larger canoe. I suspected the canoe was made about the same time as when this area of the Space Coast was being settled in the 1860s. The possibility that someone had repurposed the dugout in the past seemed reasonable as well. I could envision this artifact hanging on the wall of a restaurant or hunting and fishing lodge as a piece of decor.

There are pluses and minuses when a photo or post on Facebook or the internet goes viral. I was to experience the negative aspects as well. The kooks come out of the woodwork. I learned what "smudging" was after I had received many emails and messages insisting that I must do this immediately. Smudging is an ancient purification ritual originally practiced by Native Americans. It involves igniting a bundle or braid of dried herbs and waving the smoke to purify people, spaces, or things. Many of the messages pleaded with me to perform this ritual as soon as possible, so I did. Choosing the herb of my choice one evening, I had smoke floating over, under, and around the dugout while I spoke to it and thanked it for its service. Every passing day the dugout continued to gather steam in the media. I had done six or seven interviews on television. Channel 6 reporter Clay Lepard and I are still friends because of the dugout. I also received several calls from the native community urging me to do this or that with the dugout.

Julie Duggins in Tallahassee called me and asked if I could make the dugout available to the University of South Florida staff, which wanted to perform high-speed 3D imaging on the dugout. They wanted to have it in the shade, undercover to do this. I told them they might wish to bring a tent to set up next to the pond. They complained about the heat and sunlight being too intense. I finally told them not to worry, I would have it ready for them when they arrived. One evening under the cover of darkness, I invited a few friends over to help load and unload the dugout. Only myself, my wife Suzy, Craig, and Tom the archaeologist had seen the dugout, so the few people I'd invited to assist me were excited to see it.

We loaded it in my truck and covered it over with a tarp. The crew rode in back with the canoe, making sure they supported it the few blocks it had to go. The dugout was gently placed on two sawhorses in my garage for the evening. When I turned off the lights in the garage that evening, I looked at the dugout in the darkness with just the streetlights shining on it from the outside. It certainly had a presence, a presence which was magnified in the darkness. I'm happy to say it felt like a positive magic.

A USF professor and two associates came over the following day. I introduced myself and we exchanged pleasantries. Suzy made them a coffee cake, which they promptly refused. After trying to make conversation several times, I realized it wasn't happening. However, I noticed that as soon as I left the room, the talk picked up. I came and went during the morning and it was always the same. The

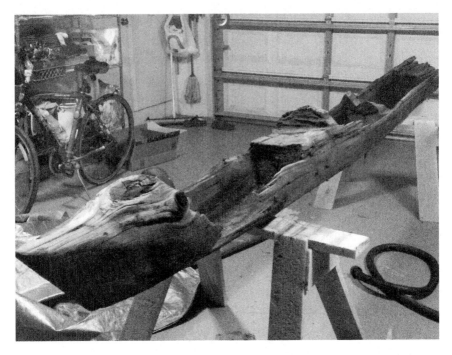

The dugout was moved to the author's garage at the request of Florida BAR authorities so that it could be scanned for 3D modeling.

little conversation I picked up on seemed to be a constant critique of associates. Plus, they were tearing up the "Ancient Aliens" guy with the big hair who hosted a UFO investigation program on television. The little social club they belonged to must have been quite exclusive if their conversation was any indication. I had set up a GoPro at the beginning to record the event, so I didn't stick around where I wasn't wanted, in my own garage. If I had it to do over again, they would have brought their own creature comforts. They were bad ambassadors, rude and narcissistic. I was satisfied to see them drive away. But before they left, they took the time to take a picture of the professor posing at the site where I showed them I found the dugout. The professor was stealing the limelight. Was I surprised(?)... not really. I would see the professor again, and she would continue with her questionable behavior. These folks have fragile egos... in a few days they posted the 3D imaging on the USF website, and it immediately got more hits than any other image on the site!

Weeks passed, and I was eager to be rid of the dugout. It had consumed just about every waking hour for the past month. The calls and emails continued to come in daily. I would refer them to the state's BAR representatives, but they would almost always call me back and tell me how nasty the state spokesperson was to them. This made me feel better oddly enough, maybe because I realized I wasn't the only person these bureaucrats held in disdain. When the state did go on record, they first and foremost let everyone know that there were heavy penalties for failure to notify them of any finds of this nature. They offered no praise, or thanks, or even a brief acknowledgment. They acted more like law enforcement officers than academicians and scholars.

A member of FPAN visited me at the request of Julie Duggins. FPAN is the acronym for Florida Public Archaeological Network. FPAN identify themselves as an NGO (non-governmental organization), but in reality, they do the bidding of the government. The organization works in conjunction with a myriad of governmental organizations. FPAN are self-centered wolves in sheep's clothing. They fought arrowhead collecting like it was the most terrible thing in the world. The FPAN rep's name was Kevin, and he initially struck me as a nice guy. He brought me over some FPAN paraphernalia, a ruler, some drink cozies, etc. I ended up showing him the dugout and where it was found. We walked up the hill and I showed him the private old graveyard where the confederate veterans slept eternally. He expressed his gratitude that I'd done the right thing as he saw it and he wanted to let me know he was nominating me for some FPAN award. After he left, I researched FPAN, and I was sorry I hadn't done that before we met.

Julie Duggins and I spoke often about the dugout, and soon we were to meet face to face. She rented a U-Haul box truck and planned to come and pick up the dugout and take it to Tallahassee. She came to my house, and we sat in my living room and completed arrangements for her to load out the dugout the following morning. Julie had a likable nature, and she was engaging. We talked for a long time, and then she shared the news that they had received the radiocarbon dating back from the lab. I was eager to hear the results, "So what was it," I asked. She said the radiocarbon dating results gave the wood a range of from 1640 to the late 1800s. She added that the dugout was not Cypress Wood, but Red Cedar. She then added that it was only the second Red Cedar canoe found in the state of Florida. Julie seemed excited about the date range, and I admit I was excited about the date range too. We scheduled a meeting on the following morning at my house and she would follow me to Craig's where the dugout was submerged in the pond behind his house. She asked if we needed any help and I told her no, we had enough folks on hand and Craig didn't wish to have a crowd of people at his house. She asked if it would be alright to bring one or two local friends. I relented and said one or two OK, but insisted that they bring no cameras. I had the photography aspects covered, and I had granted an exclusive TV interview to Clay Lepard with Channels 6 News to cover the load out. The Mayor of Cocoa had gotten in touch with me on Facebook and asked to see the dugout, which he was now calling "The Cocoa Canoe".

I told him he could see it on the loadout, so I messaged him and gave him a time to be at my house the following morning. It had been almost a month, and I was ready to see the dugout move on to any venue other than my own.

The next morning Julie arrived at my house with the box truck to load the dugout. The mayor arrived, and then there's a knock on the door, and it was the mayor's girlfriend. As we pulled out from my house, I could see a car following us. I figured it was the one or two friends of Julie who I agreed could come along. We pulled into Craig's house, and I had three or four cars behind me. Craig was not happy, and neither was I. One or two people had turned into four or five. The mayor said something about, "Even the society folks being there." Kevin from FPAN was there, and he appeared to be the culprit with the unauthorized guest. Then this chubby dude with dreads come rolling up with his Canon DSLR (camera), whom I stopped immediately.

"Who are you and no cameras," I said.

"Oh, I'm with FPAN," he replied.

"No cameras dude," I said.

"Oh come on man, I'm with FPAN, no one will see these except us. I promise, I promise," he begged.

Channel 6 was now on site, and I was pissed but I told the dude "OK, but no publishing any of those photos, you understand," I said once again.

"I promise, I promise," he said.

Just about the same time the Channel 13 news van pulls into the driveway.

"What do you want", I asked.

The cameraman replied, "A little bird told us there was a news story here."

Great, I thought, the little bird must belong to FPAN, along with the knot-headed, dread-lock dude. I told the cameraman, "Sorry, it's an exclusive story and you're on private property." He waved good-bye and walked back to his van, no problem.

Craig, the mayor, and I wrangled the dugout out of the pond while the strangers stood and looked on. The dread-head dude took pictures. I told Clay Lepard with Channel 6 to go see if Kevin with FPAN had anything to say while I was busy. Clay interviewed Kevin while Craig lifted the dugout out of the pond with a tractor. Julie wanted it rinsed off, so Craig moved it close to a spigot to hose it off. I could tell Craig was unhappy and just wished to get back to work. It was supposed to be just him, myself, and Julie, but it didn't turn out that way. I still didn't know exactly who these other folks were. They never took the time to introduce themselves. I learned later they were also members of FPAN, but one represented The Florida Historical Society and another the Cape Canaveral Cultural Center. As we began to load the dugout in the truck, the strangers suddenly came alive, jumped in the truck, laying hands on the dugout, and posing as academics for the camera. Dread-head was snapping pictures of the event. Once the dugout was in the box truck, dread-head took a group photo. The Irma Canoe was finally on its way to Tallahassee.

The next morning, I poured myself a cup of coffee and sat down to update my Facebook page with news of the dugout. I thought

I'd look at FPAN's Facebook page to see what they posted about it. The guy had lied to me about the pictures; they were posted all over FPAN's Facebook page. The real poke in the eye was there was no mention of me or Craig. They had hijacked the story. They had the poor taste to come on private property, uninvited, never introducing themselves. Now they had broken their promise about the use of the photos. They had no regard for private property or intellectual rights. They would go on to publish these photos many times, and the state museum would use one on the cover of their magazine, "Florida Frontiers". If I had that day to do over again, things would have been much different. I emailed Kevin and shared my disappointment that he had brought uninvited guests. I never heard any more about the award he was going to nominate me for... no surprise there.

Several months passed when Julie sent me an email inviting me to Tallahassee for a seminar she had arranged concerning the dugout canoe. They held the seminar at the R A Gray building in the capitol. I decided to attend and made the long drive to Tallahassee.

The capitol was quiet on that Saturday morning when I pulled into the parking lot next to the museum. I recognized a camera crew with "Florida Frontiers" walking into the building, so I followed them. They saw me. One of them sort of nodded his head, but that was the only acknowledgment I received from them. Upon entering the building, I saw a large graphic promoting the museum. On that graphic was a large picture of the gold tray we had found so many years ago. Just the sight of it made me feel sick to my stomach, especially once I knew the false narrative state reps spread regarding how the tray was found. For the past forty years, the students and guests who visited the museum were told that a father and his two sons had found the gold tray! How they came up with that one, I'll never know. There were around thirty people already sitting down when I took a seat in the theater. I saw Julie; she acknowledged me, and we spoke briefly. Julie was in charge of the seminar. She thanked everyone for coming and started the show. She spoke first and covered pretty much what everyone knew already. At this event the speakers were the specialists who worked on the Cocoa canoe; C14, Dendro, pXRF, and 3D Scanning. The first speaker talked about carbon dating. The dates on the dugout were from 1640 to the late 1800s. The next speaker talked about dendro-chronology, which is the scientific method of dating tree rings. The dugout, which we initially had thought was cypress, was actually native red cedar. Unfortunately, there is little to no data on native red cedar, so dendro-chronology didn't tell us much. One of Julie's associates then spoke about pXRF scanning which uses a field-portable/hand-held x-ray fluorescence (pXRF) analyzer which can be potentially useful for non-destructive chemical characterization of archaeological ceramics. This time instead of ceramics they used it to scan the dugout, particularly the portions with paint on them. They moved on to the 3D scanning which USF performed in my garage. The professor spoke and thanked me by name. She talked about the dugout at some length and wrapped up her talk with the possibilities it started life as a power or telephone pole. She finished her presentation with a photo of her standing next to the road where I had shown her I found the dugout. I thought that picture of her, by herself, spoke volumes about her. I didn't give this "telephone pole" theory much credence, frankly. It was like they were trying to fit a square block in a round hole. Yes, it was round like a pole, but using these criteria, almost all the dugouts in the state's collection would qualify as candidates. I paid close attention to what evidence lead them to believe it started out as a pole.

They called a break in the session, and people stretched their legs, walking around in the foyer. I was sitting down when the USF professor walked right by me, she looked at me but never spoke a word. The seminar resumed and everyone took their seats again. Julie wrapped up the seminar with a few more tidbits, but nothing new. I was waiting for some small recognition but got none. I was never introduced to the audience or any of Julie's associates. My name was only mentioned once, very briefly, by the USF professor.

They were giving away a canoe that day at a raffle, so Julie spoke about the raffle. She never said a word about me, and never mentioned my name. The lights came on in the auditorium and I got up and walked by Julie never saying a word. I felt like an idiot for ever trusting any of these people. I was so mad; I drove straight home, more on steam than on gas. Later that evening I checked my email at home and saw I had gotten an email from the USF professor, it read,

"I wanted to give you an update on our work with the canoe. There was a canoe lecture series today in Tallahassee and I presented on the 3D project with the Irma canoe. One question I had for you... how did you guys get the canoe back to your place again? When they took it from the pond, did they use a rope around it to get out or how did they remove? I don't have many images from the removal process, so if you have any, would love to see if you have a chance."

I responded, "Hi, You walked right by me three times today. I was never acknowledged being in the room. I spent 10 hours on the road today, spent $125 in gas and tolls. Should have stayed home. Your group doesn't really make one feel equal or acknowledge the time, money, and sacrifice we made for this effort. Doesn't make a person feel like it was worth it when your efforts aren't acknowledged, or your presence isn't even noted in the room. How would you feel?"

She replied to me, "Oh my God...no one told me you were there...I wish so much you would have come up and said hello. I have MS and my ability to facial recognize sometimes is terrible! I totally thanked you in my talk and you were on my recognition slide as well. This whole work comes from your involvement and from my perspective, your input and collaboration is/was crucial! Concerned citizens are so vitally important. Did anyone know you were coming? I would have thought if anyone knew you should have been personally recognized. I was totally unaware you were there, really wish you would have said hello, and feel terrible that I did not recognize you. Please accept my apologies, Randy, and I absolutely did not walk by you knowingly. You were the first person I wanted to share all with afterwards."

My last email was lengthy, so I'll only share a part of it. It read, "If I had it to do over I would let it lay with the other, 'utility poles', for the front-end loaders to pick up. They were just down the street, anyway. My integrity has been questioned, 'Was it really found where I said I found it?' 'Is it a hoax?' 'Was it really in the road or did I drag it out of the water?' The state gave me no credit, they said that it was the threat of the law and legal action that influenced my decision to call FBAR. How nice of everyone to call me out on my word. Thanks.

It's taken me away from my work, I've been referred to as a retiree, or after 40 years of photography, an amateur photographer. I've called in favors from friends to help me save, transport, secure, move, move again, and then have their property invaded by the press and interested parties trying to do the right thing. You say, 'Concerned citizens are so vitally important.' You guys are blowing it, get off your high horses. Remember the golden rule, treat people like you would wish to be treated. This has been a source of aggravation, frankly. I'd like to say I'm sorry, but I just don't feel that way, I feel like a fool for getting involved in the first place. I can see the headlines now, 'Irma Canoe turns out to be utility pole', that should all get us a merry laugh."

And with that, I never heard another word from the professor, and I didn't expect to, didn't want to, didn't need to.

Julie had emailed me and I shared the same feelings with her as I did with the professor. She sent me a letter which read as follows: "Randy, I said it before and I'll say it again, you are right. I messed up, and the state messed up - and we keep doing that by policy, almost. I'm sorry and I hope you'll accept this 3D model as a 1st step olive branch. The opening in Canaveral this fall should include you. I've resigned from B. A. R., just switching from the public sector to the private sector. Great to work with you. Julie."

I forgot to mention that in January, the Wall Street Journal did a front-page story about mysterious items that wash ashore, and the dugout and myself were on the front page. I had a business associate call me and said he spit his coffee out. He was so surprised, he congratulated me on making the front page of the Wall Street Journal without being indicted.

The dugout eventually made its way back to the Space Coast, where it was placed on display at The Cape Canaveral City Hall. The city will move it to a new city hall when it is built. Julie and an associate brought the dugout back and delivered it to Cape Canaveral. Julie invited me to speak, which I did. After I spoke, Julie and her associate took the podium and stated that their best assumption was that it was possibly made from a utility pole. This was the same nonsense that was presented by the USF professor in Tallahassee at the seminar. The reasons were many, and I will refute every one of them as follows. The dugout was Red Cedar, Eastern Red Cedar, which they claimed was a utility pole. No utility poles were made from Eastern Red Cedar. It is a gnarly, crooked tree by nature. The tree is harvested for shingles, square logs no longer than 6 feet, and pencils. When cedar is used for poles, they use Western Red Cedar, not Eastern. If you go back to the 1800s and look for Red Cedar being sold, there is very little of it. There are lots of pine tree poles, and some cypress, but no Red Cedar. I also firmly believe they did not fashion it from a tree, but a tree limb. The scholars then confused a wood lathe with a de-barker. They try to say that there are marks on the end of the dugout showing that they had turned it on a lathe. Wood mills do not turn poles made from trees on lathes to make them poles. They place the trees in a cradle-type machine called a de-barker where they are rolled, removing the bark. Let's now consider when utility/telephone poles came to be used in the area of the Space Coast. Many cities had their own power plants but not until the late 1890s as was the case with Titusville. Most cities and towns did not have any power until the early 1900s. I have looked for old photos of Brevard County and Indian River road showing poles along the road in the early 1900s and late 1800s. I can find none. The professor also attempts to associate a few marks on the canoe with metal spikes placed in poles to facilitate climbing them. They rarely used metal spikes in trees to help climb them, they were outlawed in many places. I can find no pictures of metal spikes in poles in Florida. What I find are pictures of men climbing poles with spiked boots and straps. Can you imagine the cost of installing metal spikes on all the telephone or power poles in the state? It would be cost-prohibitive and unnecessary. And to wrap it up, there is no sign that they ever buried any portion of it in the ground, as they would a pole. At the place where I found the dugout there stand two or three ancient, almost dead, Eastern Red Cedar trees, maybe cousins of the canoe. Why these trees, possibly ancestors, are ignored, I don't understand. If I had known that they would wrap up the presentation with this theory, I would have been ready for them.

Julie and her associate stopped by my house for coffee on the way back to Tallahassee. When we met I said, "You guys really know how to piss on a parade, don't you."

They acted surprised. "Why do you say that?"

"You took it from 1640 to a telephone pole. That's a big jump."

"We think it's still interesting," they said.

"I hope everyone else thinks so too, but I doubt it."

As I sit here writing, I'm thinking I need to stop by Cape Canaveral city hall and go visit the old girl, and maybe I should apologize to her.

Detail view of the bow of the dugout canoe... note the crab trap buoy blown ashore into a tree in the background.

Afterword

Memories are like little trinkets of time, you hide them away, maybe bring them out and look at them from time to time. Some trinkets lie in the bottom of your memory box, hiding from your view until you dig them out. There are some that serve you better if they remain hidden. Some deserve to be shown and shared with others. It can be a tough choice which to share and which to hideaway. Years had passed when one day I received an email from a gentleman by the name of Terry Armstrong. Terry is a writer, a diver, a salvor, and a publisher. He was writing a book about Cape Canaveral called, "West of the Bull". He wanted to ask me a few questions about his book. I was not especially nice to Terry when we first spoke. I had spent years trying to forget some chapters of my life. My efforts on the Cape wreck were a total failure, and with perceived failure comes pain. My poor attitude did not deter Terry, he coaxed a little out of me every time we talked, which we eventually did many times. Terry was kind enough to mention me and my misadventures in his books, "West of the Bull" and "The Rainbow Chasers". Terry encouraged me to tell my story and said he would publish it if I ever spent the time to write it down. The pandemic of 2020 would have most of us isolated from one another, scared of things we can't see or understand. If it wasn't for Terry Armstrong's kindness and encouragement, these pages would be blank. So I started writing, and this book is the result.

****

The last time I heard from any of the old BCC dive club members was from Mike Brady in the 1990s. Mike went to commercial dive school and was working in the Middle East. I haven't heard from him since. The body of my dive instructor and good friend, Bill Meyer with Diveco Dive Systems was eventually returned to the States after many months of negotiating with the Colombian government. The gold glove tray sits in Tallahassee, the premier piece in the State of Florida's collection of Spanish treasure. The private sector found every piece in their collection. Hindsight is 20/20 and boy golly did we ever screw up on that deal! To this day, when I beachcomb down south, I stop at the same McDonalds we stopped at the day we found that gold tray, just for luck. The state hasn't changed its ways; they are still stealing what they want from salvors under a variety of new laws. Recently they stole what will be one of the most valuable historic wrecks from Global Marine Exploration, possibly one of the French Huguenot ships that sank in 1565. For more on this read, "A Hundred Giants: New Discoveries" by Terry Armstrong.

I received terrible news in 1991 from my friend Colette Witherspoon that my friend Parker Turner was killed in a freak cave diving accident at Indian Springs because another diver released air

Standing left to right: Dan Porter, the author, and Don Porter while sitting left to right is Phil Ratcliff, Harold Holden, John Brandon, and Dave Rust.

The author on the left and Rex Stocker, circa 2021

bubbles that caused a limestone ceiling to collapse and fatally trap Parker in the cave. I had introduced Wayne Brusate and Collette to Parker soon after the *Regina* project. Wayne was trying to increase his bottom time while working the wreck and was interested in the new air mixtures which Parker and his cave diving associates had been working on for years.

Parker isn't the only one who has left us. Lou Ullian, Bill Saurwalt, Mel Fisher, Frank Allen, Sam Staples, along with Kellyco founder Stuart Auerbach are all hunting treasure in heaven. My boss at Aqua-Tech, George, passed away but I run into Skipper now and then. I haven't seen Jim Ryan, Alex Kuze, or Captain Jon Christiansen in over ten years. Tom Kraft still lives in Titusville, Florida, and was kind enough to provide many of the photos in the book. Lee Spence is writing, still exploring, still chasing the dream. Kirk Purvis, last I heard, was chasing senoritas in Panama. Roger Miklos passed away in the Florida Keys. His son was the star in a diving series on the

Left to right: Jim Ryan, the author, and Jon Christiansen, circa 2004

Discovery channel. I still talk often to my former boss at Aquatic Adventures, Bill Myer. Bill is 84 now and still in great shape. Terry Armstrong and I went to see Rex Stocker, who I hadn't seen for over 20 years. I enjoyed seeing Rex, whom I always admired.

They transformed Little Stirrup Cay from a quiet, lush tropical island into a concrete amusement park. What they did to that little cay is a crime, all in the pursuit of money. Google it, or look at it on Google Earth; it's enough to make you cry. I ran into Money and Tony Robinson many years ago when I was visiting Harbour Island in Eleuthera, and both were doing well. Mike Madden has spent much of his life exploring and mapping the caves of Mexico. He owns and operates a film production company in California and was gracious enough to provide a few photos for this book. My former wife, Penney Reese, lives in Atlanta, and works for the CDC. I would imagine she has dived little since we separated (insert shark music). Rod Steen from my Aqua-Tech days is co-owner of a successful directional boring company. I returned to Placencia, Belize, in 2000 for several weeks. Moses Leslie died, I was told when I visited. Hurricane Iris hit Placencia in 2001, destroying much of the peninsula. Places where I had stayed, were gone, not damaged, but GONE!

My mother Ruth went to be with my dad in 2002. I sense her looking down on me from time to time when I'm feeling low, telling me to "hang in there." I hope the hair perms in heaven are good ones. Captain Harry Hawkins surrendered his command of the *Miss Port Sanilac* and joined his mates on the other side. Wayne Brusate is still diving and was gracious enough to provide wonderful photos for this book. I have spoken with Sue Steinmetz, who also provided photos for this book and is doing well. She said Collette is doing well, trying to stay warm in the Michigan winters. Ed Horan left Key West and law practice with his brother Dave, moving back to Tallahassee and practicing divorce law. John Brandon is still digging, still looking, still finding... it's in his blood.

The greatest treasure I ever found was in Cocoa Beach. It came as a little lady, just over five feet tall. She grew up in Cocoa Beach, the only girl in a family of three boys. Her name is Suzy, and she tries her best to fill my days with sunshine. This can be tough, as often the clouds follow me around wherever I go. Her job isn't always easy being my partner in life, but she never gives up on me, encourages me when I need it, keeps me company on my path. Suzy gets help keeping me in good spirits from a little three-legged cat named,

"Salty Crackers" (he reminded me to mention his name). I have many people I haven't thanked, but want you to know I appreciate you. I've been beyond fortunate to have lived such "A Treasured Life". May God bless you and fill your life full of treasured memories.

Salty Crackers, Suzy and Randy Lathrop, circa 2020

329

INDEX

337

M

Madden, Mike... 233, 234, 240-250, 328
Madison... 49
Madison Blue... 9, 48
Madonna Passage... 239
mag... 152, 284
mag boat... 29, 86
mag malfunctioned... 299
magged... 291
Maggie... 166
magnetometer... 29, 152
magnetometer survey... 283
mailboxes... 33
main shipwreck...
manatees... 2
manifest... 121
many s... 221
mapping the wreck site... 133
March, Captain Guyan... 181
Marco Island... 306
marine radio... 155
Marine Sanctuary... 256
maritime heritage... 289
Mark... 171
Marley, Bob... 211
marriage... 190
Martindale, James... 277
Marx, Bob... 7, 55, 56
Marysville... 253
Maya... 213, 235
Maya Airlines... 210
Maya Blue... 244
Maya Room... 225
Mayan artifacts... 235
Mayan Blue... 236, 244
Mayan ruins... 227
Mayan sites... 228
Mayday... 156, 157
McClellanville... 105, 288
McCloskey, Mark... 49
McConkey, Captain Edward H.... 252, 253
McGehearty, Tom... 285, 286
McLarty Museum... 18
Me no walk like crab... 171
measurements... 220
Mecom, Jr., John W.... 148, 151, 154

Media Department... 197
Mel Borne's Marina... 79, 113, 115
Melbourne... 54, 113, 150, 282
Melbourne Beach... 28, 150
Melchor de Mencos... 227
Melton, Gene... 9
Melton, Mary... 9
Mennonites... 211
Menorcans... 279
mermaids... 1
Merritt Island... 43
Mesoamerica... 235
Mest, Craig... 209, 212-227, 246, 307-317
metal detector... 220, 280, 286, 300
metal spikes... 322
Meyer, Bill... 12, 325
Miami... 144
Miami... 143
Miami Beach... 151
Miami Herald... 141
Miami Vice... 142
Micco... 97
Michigan... 258
Michigan Department of Natural
Resources... 277

Michigan state archeologist... 277
Michigan's Department of Natural
Resources... 256

Mickey Mouse... 195
microfilm... 121
Mike B.... 11
Mikes Promise... 249
Miklos, Darrell... 133
Miklos, Roger... 122, 125, 133-138, 140, 326
Mini Fish sailboats... 187
Mini-14... 140, 153, 154
Minnie Mouse... 193
Minolta SLR... 246
Moe... 286
Molasses Reef... 125, 128
Mom... 246
Money... 166, 171, 188, 328
Money Path... 59, 62, 69
Money Walk... 36
Monkey Room... 49

345

About the Author

As a testament to his long diving career, Randy Lathrop holds certifications with the Professional Association Of Diving Instructors as an Open Water Instructor, and with the National Association Of Cave Diving as a Cave Diver. He's also a certified YMCA Cave Diver.

He holds a USCG 100 Ton Captain's license, and along the way, earned an Associate Science Degree in Industrial Photography. He is a National Speleological Society Cave Diving Certified Recovery Specialist and earned certifications from Scuba Schools International as an Advanced Open Water Instructor.

He has worked as a commercial diver for Quest Exploration, East Coast Research, Aqua-Tech, Circle Bar Salvage, Nomad Treasure Seekers, Cobb Coin Company, Freedom Marine, Treasure Salvors, Casper Colosimo and Sons, and Shipwrecks Inc.

As a professional photographer, he has photo credits in the Wall Street Journal, at NBC News, ABC News, the Smithsonian Magazine, Florida Sportsman Magazine, Sport Diver, Florida Today, the Washington Post, and Weather.com, to name a few. As a videographer he has had video on Inside Edition, Deep Sea Detectives, and numerous news agencies.